Springer Theses

Recognizing Outstanding Ph.D. Research

For further volumes:
http://www.springer.com/series/8790

Aims and Scope

The series "Springer Theses" brings together a selection of the very best Ph.D. theses from around the world and across the physical sciences. Nominated and endorsed by two recognized specialists, each published volume has been selected for its scientific excellence and the high impact of its contents for the pertinent field of research. For greater accessibility to non-specialists, the published versions include an extended introduction, as well as a foreword by the student's supervisor explaining the special relevance of the work for the field. As a whole, the series will provide a valuable resource both for newcomers to the research fields described, and for other scientists seeking detailed background information on special questions. Finally, it provides an accredited documentation of the valuable contributions made by today's younger generation of scientists.

Theses are accepted into the series by invited nomination only and must fulfill all of the following criteria

- They must be written in good English.
- The topic should fall within the confines of Chemistry, Physics, Earth Sciences, Engineering and related interdisciplinary fields such as Materials, Nanoscience, Chemical Engineering, Complex Systems and Biophysics.
- The work reported in the thesis must represent a significant scientific advance.
- If the thesis includes previously published material, permission to reproduce this must be gained from the respective copyright holder.
- They must have been examined and passed during the 12 months prior to nomination.
- Each thesis should include a foreword by the supervisor outlining the significance of its content.
- The theses should have a clearly defined structure including an introduction accessible to scientists not expert in that particular field.

Javier Munárriz Arrieta

Modelling of Plasmonic and Graphene Nanodevices

Doctoral Thesis accepted by
Universidad Complutense de Madrid, Spain

Author
Dr. Javier Munárriz Arrieta
Departamento de Física de Materiales
Universidad Complutense de Madrid
Madrid
Spain

Supervisors
Prof. Andrey V. Malyshev
Prof. Francisco Domínguez-Adame Acosta
Departamento de Física de Materiales
Universidad Complutense de Madrid
Madrid
Spain

ISSN 2190-5053 ISSN 2190-5061 (electronic)
ISBN 978-3-319-36150-5 ISBN 978-3-319-07088-9 (eBook)
DOI 10.1007/978-3-319-07088-9
Springer Cham Heidelberg New York Dordrecht London

Printed on acid-free paper

Springer is part of Springer Science+Business Media (www.springer.com)

To my parents

Supervisors' Foreword

In solid-state physics and material science, doped materials and compound systems have been playing the most important roles during the last decades. The idea of mixing different materials in order to benefit from a combination of their properties proved to be very fruitful. It is well known that even a relatively low concentration of dopants can change some properties of a bulk material dramatically: the vast majority of modern microelectronic devices are based on doped semiconductors. Dopants are essentially small defects diluted in a bulk material where only their averaged statistical properties matter. However, recent advances in fabrication and measurement techniques have been opening fascinating possibilities to study small systems produced in a controlled way. Modern growth methods and nanostructure technologies allow to combine various materials, exploiting their properties not only at the bare substance but also at the structural level. Nanoscopic objects such as nano particles, quantum dots or even single molecules can be produced or manipulated at the nanometer scale. Thus, these single nanoscale systems are the main protagonists now, while the bulk starts to play the secondary role of an environment. A combination of nanostructures based on different materials, making use of the most prominent properties of each constituent, is the founding idea of hybrid systems which form the backbone of this Thesis.

Unlike many other theses which typically follow one main thread, focusing on some aspects of it in detail, and therefore are quite homogeneous, the Thesis of Javier Munárriz is a quest for very different and hopefully interesting effects in totally different systems. The only characteristics that the latter have in common is that they are all hybrid nanoscale systems. Thus, transistor effect for spin-polarised currents and controlled spin filtering was predicted for nanostructures based on graphene and ferromagnetics. Nonlinear effects, such as optical and electro-optical hysteresis and bistability were predicted for hybrid systems comprising a semiconductor quantum dot in the proximity of a heterointerface with electrically tunable dielectric contrast. Negative differential resistance for spin currents was demonstrated in superlattices induced by ferromagnetic oxides deposited on top of graphene nanoribbons. Addressing metal nanoparticle-based optical antennas, several methods of formation and control of their highly directional radiation patterns were proposed.

Finally, we would like to point out that Javier Munárriz wrote the vast majority of numerical codes used for simulations of all of these different systems almost from scratch. These codes are technically complicated and very different in nature. In particular, simulations of nonlinear electro-optical effects required an implementation of methods of the electrodynamics of the stratified media. To model graphene-based devices a very efficient quantum transmission boundary method combined with the transfer matrix approach was implemented for the first time. All theoretical techniques and numerical methods are described in detail in the appendices and can readily be used, providing a solid technical background for further research work.

Madrid, April 2014

Prof. Andrey V. Malyshev
Prof. Francisco Domínguez-Adame Acosta

Contents

Symbols and Abbreviations

a_0	Interatomic distance in graphene, $a_0 \simeq 0.142\,\text{nm}$
$\tilde{a}_0$	Interatomic distance between next-nearest neighbors in graphene, $\tilde{a}_0 \equiv \sqrt{3a_0} \simeq 0.246\,\text{nm}$
aGNR	Armchair Graphene Nanoribbon
FET	Field Effect Transistor
ITO	Indium Tin Oxide
LCAO	Linear Combination of Atomic Orbitals
MNP	Metallic Nanoparticle
NDR	Negative Differential Resistance
P	Degree of polarization of the current
P_T	Degree of polarization of the transmission
QID	Quantum Interference Device
QTBM	Quantum Transmission Boundary Method
SET	Single Electron Transistor
SPP	Surface Plasmon Polariton
SQD	Semiconductor Quantum Dot
$T(E)$	Transmission probability
TB	Tight-Binding approach
$T_{\pm}(E)$	Transmission probability for electrons with spin $\sigma = \pm 1$
TMM	Transfer Matrix Method
zGNR	Zigzag Graphene Nanoribbon

Publications Related to this Thesis

Articles and Letters

1. **Toward graphene-based quantum interference devices**
   J. Munárriz, F. Domínguez-Adame, A. V. Malyshev *Nanotechnology* **22**, 365201 (2011)
2. **Graphene nanoring as a tunable source of polarized electrons**
   J. Munárriz, F. Domínguez-Adame, P. A. Orellana, A. V. Malyshev *Nanotechnology* **23**, 205202 (2012)
3. **Strong spin-dependent negative differential resistance in composite graphene superlattices**
   J. Munárriz, C. Gaul, A. V. Malyshev, P. A. Orellana, C. A. Müller, F. Domínguez-Adame *Physical Review B* **88**, 155423 (2013)
4. **Optical nanoantennas with tunable radiation patterns**
   J. Munárriz, A. V. Malyshev, V. A. Malyshev, J. Knoester *Nano Letters* **13**, 444–450 (2013)
5. **Electro-optical switching and hysteresis of hybrid systems at nanoscale**
   J. Munárriz, A. V. Malyshev *In preparation*, (2013)

Book Chapters and Proceedings

1. **Towards a graphene-based quantum interference device**
   J. Munárriz, A. V. Malyshev, F. Domínguez-Adame *GraphITA 2011*, pp. **57–60**, Springer Series on Carbon Nanostructures (2012)

Other Publications

1. **Model for hand-over-hand motion of molecular motors**
   J. Munárriz, F. Falo, J. J. Mazo *Physical Review E* **77**, 031915 (2008)
2. **Landau level shift under the influence of short-range impurities in gapless graphene**
   J. Munárriz, F. Domínguez-Adame *Journal of Physics A: Mathematical and Theoretical* **45**, 305002 (2012)

3. **Spectroscopy of the Dirac oscillator perturbed by a surface delta potential** J. Munárriz, F. Domínguez-Adame, R. P. A. Lima *Physics Letters A* **376**, 3475 (2012)

Conference Contributions

1. **Quantum interference nanoscale devices: transport properties of graphene nanorings**
   J. Munárriz, F. Domínguez-Adame, A. V. Malyshev Poster at *23rd General Conf. of the Condensed Matter Div. of the European Physical Society* Warsaw (Poland), 30 August - 3 September 2010
2. **Macroscopic tuning of broadband directional plasmonic nano-antennae**
   A. V. Malyshev, V. A. Malyshev, J. Munárriz Poster at *3rd International Topical Meeting on Nanophotonics and Metamaterials* Seefeld (Austria), 3–6 January 2011
3. **Towards graphene based quantum interference devices at nanoscale**
   J. Munárriz, F. Domínguez-Adame, A. V. Malyshev Poster at *ImagineNano 2011* Bilbao (Spain), 11–14 April 2011
4. **Toward a graphene based quantum interference device**
   F. Domínguez-Adame, J. Munárriz, A. V. Malyshev Poster at *GraphIta* Assergi-L'Aquila (Italy), 15–18 May 2011
5. **Novel design of a graphene-based quantum interference transistor**
   J. Munárriz, F. Domínguez-Adame, A. V. Malyshev Talk at *19th International Simposium of Nanostructures: Physics and Technology 2011* Ekaterinburg (Russia), 20–25 June 2011
6. **Controlling electro-optical bistability of hybrid systems at nanoscale**
   J. Munárriz, A. V. Malyshev Poster at *19th International Simposium of Nanostructures: Physics and Technology 2011* Ekaterinburg (Russia), 20–25 June 2011
7. **Electro-optical bistability and hysteresis in quantum dots near a hetero-interface**
   J. Munárriz, A. V. Malyshev Poster at *Optics of Excitons in Confined Systems XII* Paris (France), 12–16 June 2011
8. **Interferencia cuántica en anillos de grafeno**
   J. Munárriz, F. Domínguez-Adame, A. V. Malyshev Poster at *XXXIII Reunión Bienal de la Real Sociedad de Física Española* Santander (Spain), 19–23 September 2011
9. **Negative differential resistance in graphene superlattices**
   J. Munárriz, C. Gaul, F. Domínguez-Adame, P. A. Orellana, C. A. Müller, A. V. Malyshev Poster at *Graphene 2012 International Conference* Brussels (Belgium), 10–13 April 2012
10. **Spin-dependent transport in graphene superlattices**
    F. Domínguez-Adame, J. Munárriz, C. Gaul, A. V. Malyshev, C. A. Müller, P. A. Orellana Talk at *Granada'12* Granada (Spain), 9–13 September 2012

11. **Spin-dependent transport in graphene nanoribbons with a periodic array of ferromagnetic strips**
J. Munárriz, C. Gaul, A. V. Malyshev, C. A. Müller, P. A. Orellana, F. Domínguez-Adame Poster at *TNT12 International Conference* Madrid (Spain), 10–14 September 2012
12. **Electron and spin transport in graphene-based quantum devices**
F. Domínguez-Adame, J. Munárriz, C. Gaul, A. V. Malyshev, C. A. Müller, P. A. Orellana Talk at *V Congreso Nacional de Nanotecnología* Valparaíso (Chile), 1–5 October 2012

Other Conference Contributions

1. **Modelo ratchet para el movimiento hand-over-hand de motores moleculares**
J. Munárriz, F. Falo Poster at *III Congreso Nacional Insituto de Biocomputación y Física de Sistemas Complejos* Zaragoza (Spain), 1–2 March 2007
2. **Modelo ratchet para el movimiento hand-over-hand de motores moleculares**
J. Munárriz, F. Falo Poster at *XXI Sitges Conf. on Stat. Mechanics: Stat. Mechanics of Molecular Biophysics* Sietges (Spain), 2–6 June 2008
3. **Polaron dynamics at finite temperature**
F. Domínguez-Adame, A. V. Malyshev, J. Munárriz, R. P. A. Lima Invited Talk at *EPSRC Symposium Workshop on Quantum Simulations* Warwick (UK), 24–28 August 2009

Seminars Given by Invitation

1. **Filtros de espín basados en heteroestructuras de grafeno**
J. Munárriz Seminar at *Departamento de Física Fundamental. Facultad de Ciencias,* Universidad de Salamanca Salamanca (Spain), 12 March 2012
2. **Controlling radiation patterns of plasmonic nano-antennae**
J. Munárriz Seminar at *Faculty of Mathematics and Natural Sciences, University of Groningen Groningen* (The Netherlands), 23 May 2012

Chapter 1
Introduction

Due to the experimental developments and the deeper understanding of the behaviour of matter at the nanometre scale, nanoscience and nanotechnology have experienced a huge boost over the past few decades, becoming a major research topic in multiple fields of science. A non-exhaustive list of the latter includes material science, optics, chemistry, biology and biomedicine.

One of the research areas where nanoscience has had a fast and ground-breaking impact is the field of Information Technology. The constant reduction in the size of components in the semiconductor industry has smoothly led it to the nanoscale. However, most of the properties of bulk matter are greatly affected by this shrinking of the dimensions. This has led to difficult technological challenges which compromise future advances. Most of the investment is directed towards overcoming these problems without changing the base technology. However, there is also a need for the so-called *blue sky* research, that is, research which involves a change of paradigms with respect to the ones in use nowadays. It is the goal of the present work to show the results obtained in some of this new research directions.

1.1 Devices Based on Graphene

Triggered by the first successful synthesis of graphene [1], there has been an ever-growing interest in the properties and applications of this carbon material. Among its many remarkable properties, its truly two-dimensional geometry, high carrier mobility [2], large mean free path [1] and long spin-coherence lengths (up to several microns [3–5]) can be highlighted. The initial fundamental interest in its exotic properties has cleared the way to the investigation of its appealing technological applications.

Graphene has been intensively studied as a base material for energy harvesting. Its optical transmission under conditions of normal incidence is set to 97.7 % [6]. Therefore, it is a promising candidate to replace traditional transparent conducting films,

J. Munárriz Arrieta, *Modelling of Plasmonic and Graphene Nanodevices*,
Springer Theses, DOI: 10.1007/978-3-319-07088-9_1,

such as Tin-doped Indium Oxide [1] (ITO) or Aluminium-doped Zinc-Oxide (AZO), which are expensive, or conducting polymers, less efficient and toxic. Once mass production is achieved, graphene could enhance the efficiency and integration of solar cells, acting as a transparent electrode [7, 8] (for a review of the optical properties of graphene, see [6]). Furthermore, the possibility of optically exciting collective oscillations of the electron gas—plasmons—in graphene [9, 10], combined with a convenient engineering of the edges, could lead to a strong absorption enhancement of graphene flakes [11]. This could be exploited to provide atomically-thin active-harvesting materials for photovoltaic devices. Regarding energy storing, the possibility of adding large amounts of hydrogen to structures made out of graphene could in turn lead to the design of high-density, close-packed fuel cells [12–14].

Electronics is another field where graphene is called to have an important impact. Due to its outstanding electronic properties, graphene is a material of choice for devices working at radio frequencies [15–17]. Furthermore, its transparency and advantageous mechanical properties make it a suitable candidate to replace elements of electronic devices, such as touch screens and displays. This application is widely targeted as a near-term one, specially after obtaining large-scale wafers of graphene using Chemical Vapour Deposition [18, 19]. Moreover, if the different available methods to create band gaps in graphene are further developed [20–29], it could also be used to produce Field Effect Transistors (FET), taking advantage of its high thermal conductivity [30]. This promising application has been highlighted in the 2011 edition of the International Technology Roadmap for Semiconductors (ITRS) in the Emerging Research Devices section [31].

The ability of manipulating materials at increasingly smaller scales allows new devices to be envisioned, which exploit quantum phenomena as their principle of operation. It is inside this category where Single Electron Transistors (SET) can be found [32]. These transistors rely on the discretization of energy levels, which is a common feature of extremely confined systems, generally triggered by Coulomb repulsion. These levels can act as discrete channels connecting two leads. A transistor can then be engineered by setting in the system a method to externally tune their energy levels. This is usually accomplished by placing an electrostatic gate near the device. This concept has successfully been applied to graphene in [33] (for a review, see [34]). Another concept inherent to quantum mechanics is quantum interference, which can be understood in terms of the particle-wave duality. This phenomenon can be employed to engineer new electronic devices, taking designs previously applied to optical waveguides. These systems are usually termed Quantum Interference Devices (QIDs) [35]. In graphene, QIDs tuned by an external magnetic field have been experimentally and theoretically studied [36–40]. However, devices based on external magnetic fields hold little promise to be constituents of an integrated circuit. Therefore, a new QID is proposed in Chap. 3, having an electrostatic gate as its tuning element.

The possibility of developing devices which operate with individual electrons can also be extended to the control of their spin degree of freedom, thus engineering

[1] For all abbreviations see the glossary on page viii.

spintronic devices. The weak spin-orbit coupling of electrons in graphene leads to an extraordinarily long spin-coherence length [4], which allows for the coherent manipulation of this quantum degree of freedom. It has been proposed that this manipulation could be performed with adatoms [41, 42], as well as with ferromagnetic strips grown on top of it [43–46], or even using the intrinsic properties of graphene nanoribbons with zigzag edges, which may have a net magnetic moment [47, 48]. The second approach is adopted in Chaps. 4 and 5 to design devices with a response which depends on the spin of the charge carrier. More specifically, a tunable source of polarized electrons is engineered, as well as a device with spin-dependent Negative Differential Resistance (NDR).

1.2 Plasmonic Nanodevices

The enhanced control in the fabrication techniques has led to the miniaturization of a variety of electro-optical devices, e.g., antennas, optical modulators or optical storage media. However, a major drawback of this continuous process of shrinking the size of devices is the fundamental difficulty of treating light beyond the diffraction limit, which is of the order of half wavelength (hundreds of nanometres, for visible light). This prevents the integration of electro-optical systems in an efficient way.

A possible circumvention for this issue is the coupling of optical plane waves to near-field excitations via localized surface plasmons in metallic nanoparticles (MNPs; cf. [49], Chap. 6). These collective excitations of the electron gas can have, under resonant conditions, scattering cross-sections widely surpassing the geometrical cross-sections, leading to strong near-field intensities in the region surrounding the particle. Devices whose operation relies in the energy conversion between external electromagnetic waves and localized modes and vice versa have traditionally been called antennas. Therefore, the extension to visible light or, equivalently, nanometre scale, is termed *optical antennas* or *nano-antennas* [50, 51].

If embedded in a semiconductor, the field enhancement leads to an improvement in the rate of creation of electron-hole pairs [52–55]. This finds an immediate application in the production of photo-voltaic cells with a reduced amount of bulk semiconducting material. Once industry-scale is attained, this should be a milestone in the harvesting of solar energy, as the current bottleneck in the expansion of this clean energy is the cost of bulk semiconductors (for a review, see [56]).

The excitation of localized surface plasmons has also found an application in spectroscopy (for a review, see [57]). The resonance is very sensitive to the surrounding environment. Therefore, small changes of the refractive index in the surroundings lead to strong, measurable energy shifts of the resonance [58, 59]. Moreover, the strong field increases the absorption rate in molecules surrounding the MNP, as well as the spontaneous emission rate—Purcell effect [60]. This, in combination with the enhancement due to the chemical bonds formed between molecules and metal surfaces, is the reason why MNP arrays have been used to push the signal in Raman spectroscopy towards the limit of single-molecule spectroscopy [61, 62].

Similarly, in fluorescence measurements, if the plasmon resonance is tuned to match simultaneously the absorption and emission spectra, strong enhancements in the count rates are attained [63]. These techniques rely on the optimization of the device geometry to produce hot spots with strong field enhancement, which is a very active research field [64, 65].

Due to the Reciprocity Theorem, the very same concept can be applied to engineering optical antennas with custom radiation diagrams [66]. Different designs have been put forward, including log-periodic optical antennas [67], coupled gold nanorods [68] and Yagi-Uda nano-antennas [69]. Although available computational methods—generally relying on the discretization of the space [70–72]—are able to solve Maxwell equations in this complex systems and accurately predict the output of the experiments, there is a lack of simple physical models to understand and predict the properties of this systems. In Chap. 6, such a model is presented for an antenna comprising homogeneous spherical MNPs placed above an interface between two dielectrics. The interface allows the system to be excited by evanescent waves, which do not contribute to the measurements, provided the detector is in the far field region.

There are other possible mechanisms to integrate opto-electronic devices, among which Semiconductor Quantum Dots (SQDs) are a prominent example [73]. In semiconductor nanocrystals, the discretization of the energy levels is strongly size-dependent, with level spacing reaching optical frequencies for systems with sizes in the range 2–10 nm [74, 75]. In particular, core-shell ZnSe–CdSe SQDs are a material of choice, due to the availability of colloidal methods to produce these structures, as well as their stable optical properties [76].

SQDs have been widely applied in the field of in vitro biology [77]. Due to their small size, they have been used as optical labels, by functionalizing their surface to make them attach to the target molecule. Moreover, they have also been used in chemiluminescent measurements [78]. In this technique, chemical reactions within the target molecule produce transference of charge carriers to the excited states of the nanocrystals, which subsequently undergo a photoemission process. Finally, these nanostructures have been used in Photoelectrochemical Bioanalysis. Here, the optical excitation of electrons in SQDs produces a charge transfer into the biological system, which is in turn detected in a connected cathode [79].

SQDs have also been targeted as fundamental constituents of the so-called Third Generation Photovoltaic cells [80, 81]. The efficiency of previous generations is limited by the position of the band gap: the material is transparent for photons with energies below the band gap. On the other hand, electrons excited by photons with higher energy decay very rapidly to the energy of the band edge, and therefore the energy of the photon is used only partially. For the solar spectra, this leads to a maximal theoretical efficiency of 31 % for a gap placed in its optimum position, in the infrared (~1,200 nm). However, nanocrystals can be used to produce carrier multiplication, i.e., generation of multiple charge carriers from a single photon, thus raising that limit [82]. Moreover, they can also be used to produce photocurrents and therefore sensitize nanostructured solar cells in wider energy ranges [83].

The possibility of tuning the position of their energy levels, together with long decoherence times reaching the μs scale, make SQDs an ideal platform to engineer quantum computers [84]. Single-electron transport through systems controlled via external electrostatic gates has been proved and widely used (for a review, see [85]). Interacting spin qubits have also been experimentally demonstrated in this framework [86]. Due to the fact that SQDs can be optically controlled, the aforementioned properties pave the way for the design of ultrafast Quantum Memories [87–89]. In the proximity of reflecting systems, such as MNPs or interfaces, irradiated SQDs can exhibit strong nonlinear effects, which can also be applied to the design of electronic components. This possibility is explored in Chap. 7, where the bistability in the population of a SQD in close proximity to an interface between two media is studied. The self-interaction which causes this nonlinear effect can be electrically tuned by applying a back voltage. This allows electro-optical switches, modulators or optical memory cells to be envisioned using the proposed system.

1.3 Objectives and Outline

In the first part of this Thesis, graphene is presented as a base material to build transistors and spintronic devices. They take advantage of the quantum nature of charge carriers, which supersedes the usual classical description as the systems scale down. In Chap. 2, some electronic properties of graphene and graphene nanoribbons are obtained using a Tight Binding (TB) description. In addition to it, key concepts of coherent electronic transport are presented, in order to understand the following original results:

- In Chap. 3, interference effects of electrons are used to test the possibility of building a QID based on a graphene nanoring.
- In Chap. 4, the previous device is extended to provide control over the spin polarization. This is accomplished by means of a ferromagnetic insulator placed close to the arms of the ring, resulting in a tunable source of polarized electrons.
- In Chap. 5, a set of ferromagnetic strips on top of a graphene nanoribbon are shown to produce an I–V characteristic with spin-dependent NDR, which could be of great importance for non-linear electronic applications in spintronics.

In appendix A, the numerical method used to calculate the transmission through the studied samples is presented.

The second part of this Thesis deals with the integration of electro-optical systems in the nanometre scale, and comprises two main results:

- In Chap. 6, localized surface plasmons in MNPs are employed to modify the properties of incoming plane waves and engineer radiating patterns, thus building up a nano-antenna.
- In Chap. 7, the dynamics of a SQD in close proximity to an interface is studied. It is shown that a back gate allows the inner state of the SQD to be controlled.

The two aforementioned results are accompanied by the numerical method in which they rely, which calculates the effect of interfaces in the emission of neighbouring systems, accomplished using Sommerfeld integrals. This is detailed in appendix B.

Finally, the main results and conclusions will be summarized, and some new possible research directions motivated by the present work. Note that a summary with the main points of this Thesis is also available in Spanish, before the appendices.

References

1. K.S. Novoselov et al., Electric field effect in atomically thin carbon films. Science **306**, 666–669 (2004)
2. K.I. Bolotin et al., Temperature-dependent transport in suspended graphene. Phys. Rev. Lett. **101**, 096802 (2008)
3. C.L. Kane, E.J. Mele, Quantum spin hall effect in graphene. Phys. Rev. Lett. **95**, 226801 (2005)
4. N. Tombros et al., Electronic spin transport and spin precession in single graphene layers at room temperature. Nature **448**, 571–574 (2007)
5. O.V. Yazyev, Hyperfine interactions in graphene and related carbon nanostructures. Nano Lett. **8**, 1011–1015 (2008)
6. F. Bonaccorso et al., Graphene photonics and optoelectronics. Nat. Photonics **4**, 611–622 (2010)
7. X. Wang et al., Transparent, conductive graphene electrodes for dye-sensitized solar cells. Nano Lett. **8**, 323–327 (2008)
8. X. Li et al., Transfer of large-area graphene films for high-performance transparent conductive electrodes. Nano Lett. **9**, 4359–4363 (2009)
9. J. Chen et al., Optical nano-imaging of gate-tunable graphene plasmons. Nature **487**, 77–81 (2012)
10. Z. Fei et al., Gate-tuning of graphene plasmons revealed by infrared nano-imaging. Nature **487**, 82–85 (2012)
11. S. Thongrattanasiri et al., Complete optical absorption in periodically patterned graphene. Phys. Rev. Lett. **108**, 047401 (2012)
12. D.C. Elias et al., Control of graphene's properties by reversible hydrogenation: evidence for graphane. Science **323**, 610–613 (2009)
13. S. Patchkovskii et al., Graphene nanostructures as tunable storage media for molecular hydrogen. Proc. Nat. Acad. Sci. U.S.A. **102**, 10439–10444 (2005)
14. G.K. Dimitrakakis et al., Pillared graphene: a new 3-D network nanostructure for enhanced hydrogen storage. Nano Lett. **8**, 3166–3170 (2008)
15. F. Schwierz, Graphene transistors. Nat. Nanotechnol. **5**, 487–496 (2010)
16. Y. Wu et al., High-frequency, scaled graphene transistors on diamond-like carbon. Nature **472**, 74–78 (2011)
17. S.-J. Han et al., High-frequency graphene voltage amplifier. Nano Lett. **11**, 3690–3693 (2011)
18. X. Li et al., Large-area synthesis of high-quality and uniform graphene films on copper foils. Science **324**, 1312–1314 (2009)
19. S. Bae et al., Roll-to-roll production of 30-inch graphene films for transparent electrodes. Nat. Nanotechnol. **5**, 574–578 (2010)
20. F. Guinea et al., Electronic states and Landau levels in graphene stacks. Phys. Rev. B **73**, 245426 (2006)
21. B. Partoens, F.M. Peeters, From graphene to graphite: electronic structure around the K point. Phys. Rev. B **74**, 075404 (2006)
22. Y. Zhang et al., Direct observation of a widely tunable bandgap in bilayer graphene. Nature **459**, 820–823 (2009)

23. R. Balog et al., Bandgap opening in graphene induced by patterned hydrogen adsorption. Nat. Mater. **9**, 315–319 (2010)
24. J.G. Pedersen, T.G. Pedersen, Band gaps in graphene via periodic electrostatic gating. Phys. Rev. B **85**, 235432 (2012)
25. F. Guinea et al., Energy gaps and a zero-field quantum hall effect in graphene by strain engineering. Nat. Phys. **6**, 30–33 (2010)
26. S.-M. Choi et al., Effects of strain on electronic properties of graphene. Phys. Rev. B **81**, 081407 (2010)
27. M. Kim et al., Fabrication and characterization of large-area, semiconducting nanoperforated graphene materials. Nano Lett. **10**, 1125–1131 (2010)
28. L. Yang et al., Quasiparticle energies and band gaps in graphene nanoribbons. Phys. Rev. Lett. **99**, 186801 (2007)
29. M.Y. Han et al., Energy band-gap engineering of graphene nanoribbons. Phys. Rev. Lett. **98**, 206805 (2007)
30. A.A. Balandin et al., Superior thermal conductivity of single-layer graphene. Nano Lett. **8**, 902–907 (2008)
31. The International Technology Roadmap for Semiconductors (ITRS). Semiconductor Industry Association (2011)
32. N. Ubbelohde et al., Measurement of finite-frequency current statistics in a single-electron transistor. Nat. Commun. **3**, 612 (2012)
33. L.-J. Wang et al., A graphene quantum dot with a single electron transistor as an integrated charge sensor. Appl. Phys. Lett. **97**, 262113 (2010)
34. T. Ihn et al., Graphene single-electron transistors. Mater. Today **13**, 44–50 (2010)
35. C.A. Stafford et al., The quantum interference effect transistor. Nanotechnology **18**, 424014 (2007)
36. S. Russo et al., Observation of Aharonov-Bohm conductance oscillations in a graphene ring. Phys. Rev. B **77**, 085413 (2008)
37. M. Zarenia et al., Simplified model for the energy levels of quantum rings in single layer and bilayer graphene. Phys. Rev. B **81**, 045431 (2010)
38. J. Wurm et al., Graphene rings in magnetic fields: Aharonov-Bohm effect and valley splitting. Semicond. Sci. Technol. **25**, 034003 (2010)
39. Z. Wu et al., Quantum tunneling through graphene nanorings. Nanotechnology **21**, 185201 (2010)
40. J. Schelter et al., Interplay of the Aharonov-Bohm effect and Klein tunneling in graphene. Phys. Rev. B **81**, 195441 (2010)
41. A.H. Castro Neto, F. Guinea, Impurity-induced spin-orbit coupling in graphene. Phys. Rev. Lett. **103**, 026804 (2009)
42. D. Soriano et al., Hydrogenated graphene nanoribbons for spintronics. Phys. Rev. B **81**, 165409 (2010)
43. H. Haugen et al., Spin transport in proximity-induced ferromagnetic graphene. Phys. Rev. B **77**, 115406 (2008)
44. J. Zou et al., Negative tunnel magnetoresistance and spin transport in ferromagnetic graphene junctions. J. Phys.: Condens. Matter **21**, 126001 (2009)
45. Y. Gu et al., Equilibrium spin current in ferromagnetic graphene junction. J. Appl. Phys. **105**, 103711 (2009)
46. J. Maassen et al., Graphene spintronics: the role of ferromagnetic electrodes. Nano Lett. **11**, 151–155 (2011)
47. Y.-W. Son et al., Half-metallic graphene nanoribbons. Nature **444**, 347–349 (2006)
48. Z.P. Niu, D.Y. Xing, Spin filter effect and large magnetoresistance in the zigzag graphene nanoribbons. Eur. Phys. J. B **143**, 139–143 (2010)
49. S.A. Maier, H.A. Atwater, Plasmonics: localization and guiding of electromagnetic energy in metal/dielectric structures. J. Appl. Phys. **98**, 011101 (2005)
50. P. Bharadwaj et al., Optical antennas. Adv. Opti. Photonics **1**, 438–483 (2009)
51. L. Novotny, N. van Hulst, Antennas for light. Nat. Photonics **5**, 83–90 (2011)

52. H. Tan et al., Plasmonic light trapping in thin-film silicon solar cells with improved self-assembled silver nanoparticles. Nano Lett. **12**, 4070–4076 (2012)
53. F.J. Beck et al., Tunable light trapping for solar cells using localized surface plasmons. J. Appl. Phys. **105**, 114310 (2009)
54. B.P. Rand et al., Long-range absorption enhancement in organic tandem thin- film solar cells containing silver nanoclusters. J. Appl. Phys. **96**, 7519–7526 (2004)
55. R.B. Konda et al., Surface plasmon excitation via Au nanoparticles in n-CdSe/p-Si heterojunction diodes. Appl. Phys. Lett. **91**, 191111 (2007)
56. H.A. Atwater, A. Polman, Plasmonics for improved photovoltaic devices. Nat. Mater. **9**, 205–213 (2010)
57. K.A. Willets, R.P. Van Duyne, Localized surface plasmon resonance spectroscopy and sensing. Annu. Rev. Phys. Chem. **58**, 267–297 (2007)
58. G. Raschke et al., Biomolecular recognition based on single gold nanoparticle light scattering. Nano Lett. **3**, 935–938 (2003)
59. A.D. McFarland, R.P. Van Duyne, Single silver nanoparticles as real-time optical sensors with zeptomole sensitivity. Nano Lett. **3**, 1057–1062 (2003)
60. E. Waks, D. Sridharan, Cavity QED treatment of interactions between a metal nanoparticle and a dipole emitter. Phys. Rev. A **82**, 043845 (2010)
61. S. Nie, S.R. Emory, Probing single molecules and single nanoparticles by surface-enhanced Raman scattering. Science **275**, 1102–1106 (1997)
62. A.D. McFarland et al., Wavelength-scanned surface-enhanced Raman excitation spectroscopy. J. Phys. Chem. B **109**, 11279–11285 (2005)
63. A. Kinkhabwala et al., Large single-molecule fluorescence enhancements produced by a bowtie nanoantenna. Nat. Photonics **3**, 654–657 (2009)
64. C. Höoppener et al., Self-similar gold-nanoparticle antennas for a cascaded enhancement of the optical field. Phys. Rev. Lett. **109**, 017402 (2012)
65. H.-J. Chang et al., A photonic-crystal optical antenna for extremely large local- field enhancement. Opt. Express **18**, 24163–24177 (2010)
66. P. Biagioni et al., Nanoantennas for visible and infrared radiation. Rep. Prog. Phys. **75**, 024402 (2012)
67. R.S. Pavlov et al., Log-periodic optical antennas with broadband directivity. Opt. Commun. **285**, 3334–3340 (2012)
68. A.I. Denisyuk et al., Transmitting hertzian optical nanoantenna with free-electron feed. Nano Lett. **10**, 3250–3252 (2010)
69. T. Coenen et al., Directional emission from plasmonic Yagi-Uda antennas probed by angle-resolved cathodoluminescence spectroscopy. Nano Lett. **11**, 3779–3784 (2011)
70. F.J. García de Abajo, A. Howie, Retarded field calculation of electron energy loss in inhomogeneous dielectrics. Phys. Rev. B **65**, 115418 (2002)
71. K. Yee, Numerical solution of initial boundary value problems involving maxwell's equations in isotropic media. IEEE Trans. Antennas Propag. **14**, 302–307 (1966)
72. A.F. Oskooi et al., Meep: a flexible free-software package for electromagnetic simulations by the FDTD method. Comput. Phys. Commun. **181**, 687–702 (2010)
73. A.D. Yoffe, Semiconductor quantum dots and related systems: electronic, optical, luminescence and related properties of low dimensional systems. Adv. Phys. **50**, 1–208 (2001)
74. A.I. Ekimov et al., Absorption and intensity-dependent photoluminescence measurements on CdSe quantum dots: assignment of the first electronic transitions. J. Opt. Soc. Am. B: Opt. Phys. **10**, 100–107 (1993)
75. D.J. Norris, M.G. Bawendi, Measurement and assignment of the sizedependent optical spectrum in CdSe quantum dots. Phys. Rev. B **53**, 16338–16346 (1996)
76. S. Kalele et al., Nanoshell particles: synthesis, properties and applications. Curr. Sci. **91**, 1038–1052 (2006)
77. R. Gill et al., Semiconductor quantum dots for bioanalysis. Angew. Chem. Int. Ed. **47**, 7602–7625 (2008)

78. A. Roda et al., Bio- and chemiluminescence in bioanalysis. Fresenius J. Anal. Chem. **366**, 752–759 (2000)
79. L. Sheeney-Haj-Ichia et al., Efficient generation of photocurrents by using CdS/carbon nanotube assemblies on electrodes. Angew. Chem. Int. Ed. **44**, 78–83 (2005)
80. S. Rüuhle et al., Quantum-dot-sensitized solar cells. ChemPhysChem **11**, 2290–2304 (2010)
81. A.J. Nozik, Nanoscience and nanostructures for photovoltaics and solar fuels. Nano Lett. **10**, 2735–2741 (2010)
82. R.D. Schaller, V.I. Klimov, High efficiency carrier multiplication in PbSe nanocrystals: implications for solar energy conversion. Phys. Rev. Lett. **92**, 186601 (2004)
83. A. Kongkanand et al., Quantum dot solar cells. Tuning photoresponse through size and shape control of CdSe-TiO_2 architecture. J. Am. Chem. Soc. **130**, 4007–4015 (2008)
84. D. Loss, D.P. DiVincenzo, Quantum computation with quantum dots. Phys. Rev. A **57**, 120–126 (1998)
85. W.G. van der Wiel et al., Electron transport through double quantum dots. Rev. Mod. Phys. **75**, 1–22 (2002)
86. J.R. Petta et al., Coherent manipulation of coupled electron spins in semiconductor quantum dots. Science **309**, 2180–2184 (2005)
87. C. Simon et al., Quantum memories. Eur. Phys. J. D **58**, 1–22 (2010)
88. M. Kroutvar et al., Optically programmable electron spin memory using semiconductor quantum dots. Nature **432**, 81–84 (2004)
89. R.J. Young et al., Single electron-spin memory with a semiconductor quantum dot. New J. Phys. **9**, 365 (2007)

Part I
Electronic Nanodevices Based on Graphene

Chapter 2
Tight-Binding Description of Graphene Nanostructures

In graphene, the four electrons in the outer shell of carbon atoms are arranged in a planar hybridization sp_2, with three orbitals oriented towards the vertices of a regular triangle. Three electrons form covalent bonds with the neighbouring atoms, thus building a hexagonal lattice. The remaining electron, corresponding to the non-hybridized p orbital perpendicular to the structure, is responsible for the conductivity of the system. This is due to the non-negligible overlap between the orbitals of neighbouring atoms, which allows the electron to form extended states spanning over multiple sites.

2.1 Dispersion Relation of Graphene

The electronic properties of graphene can be described using a simple TB model [1]. The electrons in the covalent bonds form deep fully filled valence bands, and thus their effects on the conductivity can be safely disregarded. The unhybridized p orbital is only slightly perturbed by the neighbouring atoms. Therefore, the wave function of an electron in the system can be written as a Linear Combination of Atomic Orbitals (LCAO). Using these orbitals as the basis set to represent the wave function, the Hamiltonian that governs the dynamics of the electron is given by:

$$\mathcal{H} = \sum_i \epsilon_i \left|\phi_i\right\rangle \left\langle\phi_i\right| - \sum_l \sum_{\{\langle i,j\rangle\}_l} t_l \left(\left|\phi_i\right\rangle \left\langle\phi_j\right| + \left|\phi_j\right\rangle \left\langle\phi_i\right|\right), \tag{2.1}$$

where ϵ_i represents the onsite energy at the ith atom, $|\phi_i\rangle$ the atomic orbital in the same site, $\{\langle i, j\rangle\}_l$ the set of couples of lth-nearest neighbours, and t_l the hopping parameter between them that represents the overlap between orbitals. The number of neighbours included in the calculation depends on the required accuracy, and usually ranges from 1 to 3. For brevity, t_1 is redefined as t.

J. Munárriz Arrieta, *Modelling of Plasmonic and Graphene Nanodevices*,
Springer Theses, DOI: 10.1007/978-3-319-07088-9_2,

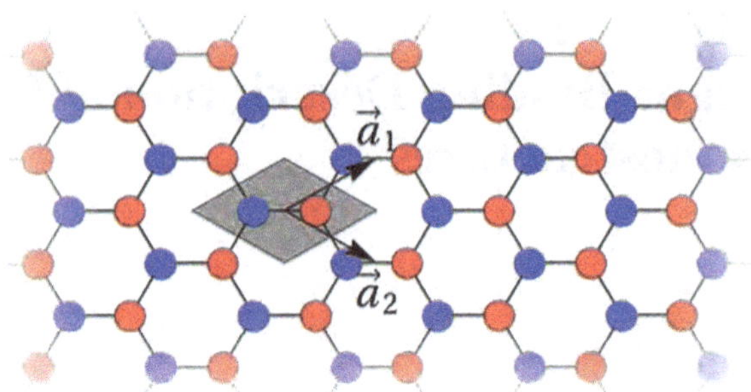

Fig. 2.1 Characteristic hexagonal lattice of graphene. Different colours are used for atoms belonging to different sublattices. A primitive cell is shown in *grey* along with the lattice vectors a_1 and a_2

If the system is assumed to be infinite, and all the onsite energies are equal (system without disorder), then the Bloch theorem can be applied. As a consequence, extended states with a well-defined **k** vector can be defined, with a wave function given by the following *ansatz*:

$$|\psi_B(\mathbf{k})\rangle = \sum_{m,n}\left[|\phi^a_{m,n}\rangle + c\left|\phi^b_{m,n}\right\rangle\right]e^{i\mathbf{k}\cdot\mathbf{R}_{m,n}} \tag{2.2}$$

where c is a complex variable, $\mathbf{R}_{m,n} = m\mathbf{a}_1 + n\mathbf{a}_2$ is the position of the centre of the $\{m, n\}$-cell and $|\phi^{a(b)}_{m,n}\rangle$ represents the atomic orbital on the first (second) atom of that cell (see Fig. 2.1). Both the arbitrary global phase of the wave function and its normalization have been used to remove the coefficient associated to $|\phi^a_{m,n}\rangle$. Plugging this wave function in the usual eigenvalue equation $\mathcal{H}|\psi_B\rangle = E|\psi_B\rangle$ and projecting it over $\left\langle\phi^a_{0,0}\right|, \left\langle\phi^b_{0,0}\right|$, leads to a system of two equations and two unknowns, E and c. This allows for the writing of the dispersion relation $E(k_x, k_y)$. The analytic expressions derived can be written in a compact way for $l = 1, 2$ [2]. In the case $l = 1$,

$$E(\mathbf{k}) = \pm t\sqrt{3 + 2\cos\left(\sqrt{3}k_y a_0\right) + 4\cos\left(\frac{\sqrt{3}}{2}k_y a_0\right)\cos\left(\frac{3}{2}k_x a_0\right)} \tag{2.3}$$

$$c(\mathbf{k}) = E(\mathbf{k})\left[e^{ik_x a_0} + 2e^{-ik_x a_0/2}\cos\left(\sqrt{3}k_y a_0/2\right)\right]^{-1}, \tag{2.4}$$

$a_0 \simeq 0.142$ nm being the distance between neighbouring carbon atoms in graphene.

Using $t_2 = -0.2t$, $t_3 = 0.025t$, as obtained in [3] using *ab-initio* calculations, the dispersion relation was calculated, and plotted in Fig. 2.2. As the system is modelled using two orbitals per unit cell, two energy bands show up in the dispersion relation. Contrary to the expression derived in Eq. (2.3), in this case there is strong asymmetry between the bands due to the non-nearest neighbour interactions.

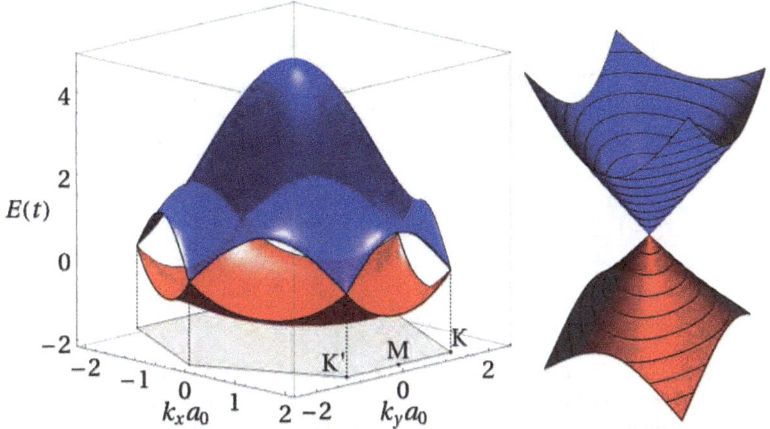

Fig. 2.2 Dispersion relation of graphene in the first Brillouin zone, calculated using a TB model considering up to third-nearest neighbours interaction and hopping parameters t, $t_2 = -0.2t$, $t_3 = 0.025t$. Six Dirac points are observed, with only two of them—K, K'—being inequivalent. A zoom in one of the Dirac points is shown in the *right plot*, where the linear dispersion relation can be observed

2.1.1 Dirac Points and Dirac Cones

The calculated bands touch each other at six points, located in the vertices of the hexagon corresponding to the first Brillouin zone. Due to the boundary conditions in the reciprocal space, only two of them are non-equivalent. They are denoted as K, K'. Contrary to the usual parabolic shape, the dispersion relation near the band edges is linear, as seen in the zoom around K, in the right part of the same figure.

These points are called Dirac points, and the neighbouring regions, Dirac cones. This nomenclature has its roots in the field of theoretical physics, where linear dispersion relations arose as solutions of the Dirac equation for massless fermions. In fact, for each cone a massless Dirac Hamiltonian can be derived as an effective Hamiltonian for states close to the Dirac point [4]. This effective description is used in Chap. 5. Interestingly, due to the presence of two atoms per unit cell, a new variable appears which is analogous to the spin of a fermion, thus being called pseudospin. Some unusual effects can be explained by means of this effective description, such as the Klein tunnelling [5], i.e., the perfect transmission of a particle through electrostatic barriers of arbitrary height and width, for normal incidence. The anomalous integer quantum Hall effect, i.e., the shift by 1/2 in the position of the Hall plateaus with respect to the usual sequence, due to the presence of a Landau level at $E = 0$, can also be explained using the Dirac equation.

The physical nature of the pseudospin comes from Eq. (2.4), which is a bivalued function with constant absolute value $|c(\mathbf{k})| = 1$. This function has branch points in the edges of the Brillouin zones, and therefore a closed loop around one of these

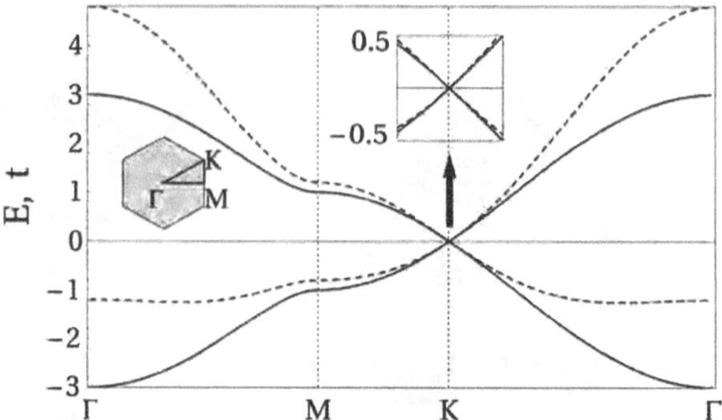

Fig. 2.3 Comparison of the dispersion relation of graphene along the path Γ–M–K–Γ using two different sets of TB model parameters: a simple nearest-neighbour with hopping parameter t—*solid line*, and another with interaction up to third-nearest neighbours and parameters t, $t_2 = -0.2t$, $t_3 = 0.025t$—*dashed line*. In spite of the strong shifts of the bands, the simplest model captures to a very high accuracy the behaviour close to the neutrality point, with both models showing the Dirac point and the linear dispersion relation

points introduces a change of sign in $c(\mathbf{k})$, which is analogous to the behaviour of the spin in fermions under 2π spatial rotations.

This analogy with massless fermions is of practical importance if the linear dispersion is stable, in contrast to being just a mathematical consequence of an over-simplified model considering only nearest-neighbour interactions. Considering interaction up to third-nearest neighbours, the expansion of the dispersion relation around the Dirac point K can be made, using $\mathbf{k} = \mathbf{K} + k_\rho \mathbf{u}_\theta$:

$$E(k_\rho, \theta) = (3t_2 - \epsilon_0) \pm \frac{3}{2}(t_1 - 2t_3)\,k_\rho - \frac{18t_2 \pm 3\sin(3\theta)\,(t_1 + 4t_3)}{8}k_\rho^2 + \mathcal{O}(k_\rho^3), \tag{2.5}$$

where the different signs stand for the two bands. The expansion around K' only changes the sign of θ. The linear term proofs the stability of the Dirac cone centred in $\mathbf{K}$. It is also clear from Eq. (2.5) that t_2 shifts the energy spectrum and t_3 changes the slope of the dispersion relation, but no angular dependence of the spectrum is observed until terms of order k_ρ^2. This effect is named trigonal warping and, as it is proportional to $t_1 \gg t_2, t_3$, is the main non-linear effect to be taken into account as the energies of interest move away from the Dirac point [6] (Fig. 2.3).

2.1.2 Opening Gaps in Graphene

Despite its impressive properties, graphene is not well suited for applications in digital electronics due to two closely related properties: the absence of a band gap and the difficulties to confine carriers using electrostatic potentials.

A number of methods have been proposed in the literature to circumvent these issues:

- Stacks of graphene layers [7–9]: placing graphene layers in stacks, with optional back gate voltages, can induce band gaps in the resulting system.
- Chemical doping: the adsorption of molecules by the p orbitals in a patterned way, e.g., by using Moiré patterns caused by the substrate [10–12], allows for the formation of heterostructures with band gaps.
- Strain induced band gaps [13, 14] and patterned defects [15].
- Band gaps induced by lateral confinement [16, 17]: graphene strips have a band gap due to the lateral confinement of the charge carriers, which increases as the system is narrowed down.

In Chaps. 3–5 the latter method is used to create devices with band gaps.

2.2 Electronic Properties of Graphene Nanoribbons

Due to the importance of opening a gap in graphene for many applications, a number of research groups are involved in the production of nanoribbons. There are currently a variety of techniques which allow for atomic precision in graphene growth, which range from the initial attempts using lithographic methods [17], to the bottom-up growth via cyclodehydrogenation after surface-assisted coupling of molecular precursors into linear polyphenylenes [18, 19] or the unzipping of carbon nanotubes [20, 21].

Despite the finite size of any sample, a useful method to understand its electronic properties is to consider one of the dimensions of the sample of infinite length, keeping the other finite. Then, the system becomes quasi one-dimensional, considering unit cells with length equal to the periodicity of the structure and same width as the structure. Then, the electronic transport can be studied using the dispersion relation $E(k)$, which can be obtained using a modified Transfer Matrix Method (TMM) [22]. Further details of this method are given in appendix A.

2.2.1 Relationship Between Dispersions of the 1-D and 2-D Systems

It is interesting to understand the results obtained for the nanoribbons in terms of the properties of bulk graphene. There is a fundamental rule connecting both systems, which states that the propagating eigenmodes of a one-dimensional system can be written as a linear combination of the eigenmodes of its two-dimensional counterpart, provided that complete primitive cells are used to construct the 1-D system. This is valid as long as no edge deformation is taken into account. Within the TB model, the boundary conditions of one cell with $n = 1, \ldots, N$ rows of atoms are equivalent to

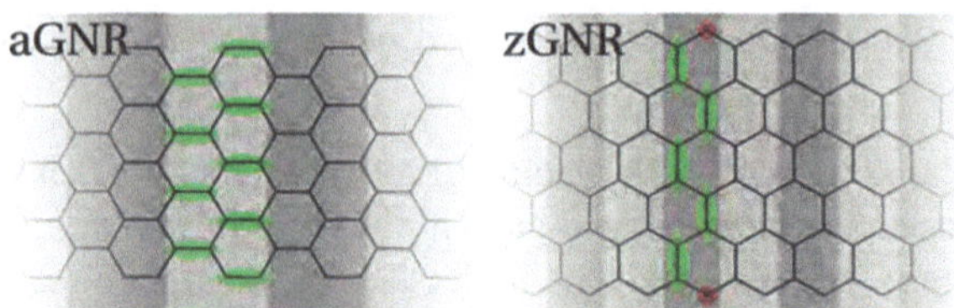

Fig. 2.4 *Left panel* shows the unit cell of an armchair graphene nanoribbon ($N = 4$ cells wide) and *right panel*, a zigzag graphene nanoribbon ($N = 3$ cells wide). The primitive cells of these structures are marked with alternate *grey tones*, the primitive cells of bulk graphene, in *green*, and the unpaired carbon atoms, in *red*

setting the wave function to be 0 at $n = 0, N + 1$. Then, a wave function in the 1-D system can be directly connected to a set of wave functions in the 2-D system.

The discussion above can be easily applied to the two types of graphene nanoribbons, with armchair edges—aGNRs, left plot of Fig. 2.4—and with zigzag edges—zGNRs, right plot in the same figure. The confinement in the vertical direction creates a new periodicity in both systems, $a = 3a_0$ for aGNRs and $a = \sqrt{3}a_0$ for zGNRs. The primitive cells of these systems are marked with alternate grey tones. For aGNRs, this cell is made up of primitive cells of the 2-D system, marked in green, while for zGNRs, two atoms—in red—are unpaired. Therefore, for aGNRs,

$$\mathbf{e}_n^{1\mathrm{D}}(k_x, E) = \int dk_y \; c_n(k_x, k_y, E)\mathbf{e}^{2\mathrm{D}}(k_x, k_y, E), \tag{2.6}$$

that is, the nth propagating eigenmode $\mathbf{e}_n^{1\mathrm{D}}(k_x, E)$ can be written as a linear combination of the eigenmodes of the two-dimensional one, as they are a complete basis set for the aGNR. On the other hand, for zGNRs, the above equation does not hold.

These arguments are tested in Fig. 2.5, by projecting the dispersion relation of graphene in k_x—aGNRs, red—and k_y—zGNRs, blue. For ribbons with armchair edges, the comparison with the 1-D dispersion—black lines in the left panel–shows a perfect agreement, contrary to the case of zGNRs—right panel. In both cases, the width of the GNR was chosen to be $N = 23$ cells.

Some additional properties can be derived from this procedure. First, the projection always mixes states from Dirac cones. In aGNRs, these cones are inequivalent, i.e., not connected by reciprocal vectors, whereas for zGNRs, the mixed cones are equivalent. Furthermore, the eigenmodes for aGNRs can be directly connected to lines of the 2-D dispersion with constant k_y; the reflections in the boundary of the Brillouin zone are due to the shrinking of the zone by a factor 2.

2.2.2 Dependence of the Dispersion Relation on the Model Used

Using the simplest TB model with nearest-neighbour interactions, very different dispersion relations can be obtained, depending on the geometry of the system [23]. If zGNRs are chosen (black lines in the right plot of Fig. 2.5), no band gap shows

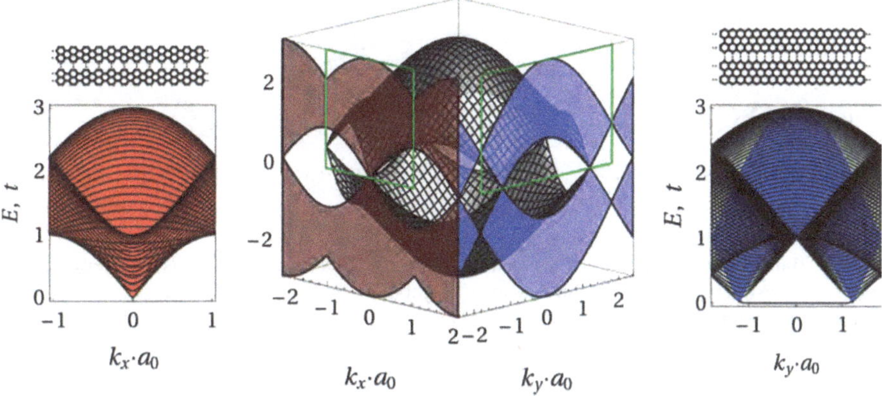

Fig. 2.5 *Central panel* in *grey*, dispersion relation of graphene in the first Brillouin zone, calculated using a TB model with nearest neighbours interaction. The projection of all the possible energy values for a constant k_x (k_y) are plotted in *red* (*blue*) on the *same panel*, including the contribution of the different Brillouin zones. *Left panel* projection of the dispersion relation of graphene—*red region*—and dispersion relation of an aGNR $N = 23$ cells wide—*black lines*. *Right panel* projection of the dispersion relation of graphene—*blue region*—and dispersion relation of a zGNR $N = 23$ cells wide—*black lines*

up, and the dispersion relation presents edge states (extended states fading away far from the edge) with energy $E = 0$.

For aGNRs (left plot of Fig. 2.5), the dispersion relations are centred around $k_x = 0$, and their band edges are in general parabolic. An important exception occurs for widths $N = 3n - 1$, $n \in \mathbb{N}$, where a gapless mode with linear dispersion appears, whereas for $N \neq 3n - 1$, a gap opens, as in the case plotted in the figure.

Nevertheless, if more elaborated models are used, the properties of the dispersion relation are modified [3]. In the case of aGNRs, the three different families persist, defined by N modulo 3, but the gapless modes fade away, and the lower edge of the dispersion relation becomes parabolic. This is shown in Fig. 2.6, where the band gap is plotted as a function of the width for models with increasing complexity: only nearest neighbours—left, interactions up to third-nearest neighbours—centre, and after the addition of hydrogenic impurities that relaxate the edges—right.

In spite of the differences between families, the band gap of each of them approximately depends on the width as N^{-1}. This is a consequence of the confinement energy of a Dirac fermion in an infinite well being inversely proportional to the width of the well [24]. This dependence has been obtained using first—principles calculations [16] and experimentally verified [17]. Note that this behaviour is different from the usual N^{-2} observed in other lattices, which stems from the solution of the Schrödinger equation for a massive particle confined in a well.

Changes in the dispersion relation of zGNRs due to the different models are not addressed here. This is due to the controversy regarding fundamental aspects such as edge reconstruction [25], or the presence of a net magnetic moment, which would make the dispersion relation spin-dependent [26, 27].

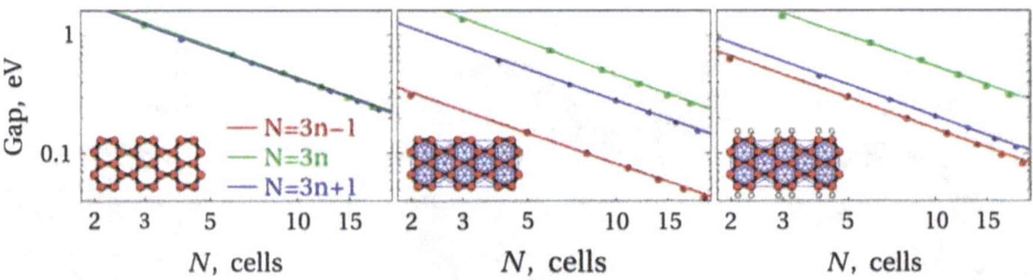

Fig. 2.6 Band gap as a function of the system width for aGNRs, calculated using three different models: nearest neighbour—*left*, third-nearest neighbours—*middle*, and third-nearest neighbours with hydrogenation in the edges—*right*

2.3 Electronic Transport Through Quantum 1-D Systems

The Landauer-Büttiker scattering formalism [28–30] has been used to calculate transport properties of the devices proposed in this work. Below, a brief summary of this formalism is given.

First, an infinite 1-D system with N propagating channels without scattering, that is, with perfect transmission, is considered, each channel having an associated dispersion relation, $E_n(k)$. The density of states of each mode n per unit of length and energy is given by

$$\rho_n(E) = \frac{2}{2\pi}\frac{1}{\partial E_n/\partial k} = \frac{1}{\pi\hbar}\frac{1}{v_{g,n}(E)}, \tag{2.7}$$

where $v_{g,n}(E)$ is the group velocity of the nth mode. The degeneracy due to the spin has already been included. Then, the current density per unit of energy associated to this mode can be expressed as:

$$j_{n,0} = e\,\rho_n(E)\,v_{g,n}(E) = \frac{e}{\pi\hbar} = \frac{2e}{h}, \tag{2.8}$$

which is independent of the dispersion relation. If the number of propagating channels at energy E is $N(E)$, the total current flowing from left to right in that perfect sample can be written as:

$$I_0 = \frac{2e}{h}\int N(E)\,[\,f_l(E, V_{SD}) - f_r(E, V_{SD})\,]\,dE, \tag{2.9}$$

where the Fermi functions of the left and right contacts are given by $f_l(E, V_{SD}) = \left[1+\exp\left(E_F - E\right)/k_B T\right]^{-1}$ and $f_r(E, V_{SD}) = [1 + \exp(E_F - eV_{SD} - E)/k_B T]^{-1}$ respectively, V_{SD} is the source-drain voltage applied across the whole sample in the x direction.

Now, instead of the infinite homogeneous system, a sample between two semi-infinite homogeneous leads is considered. Scattering within the sample makes the current density per mode j_n drop, due to the back-scattered waves. Therefore, by defining the transmission coefficient as

$$T(E) = \sum_n \frac{j_n}{j_{n,0}} \equiv \sum_n T_n(E) \leq N(E), \tag{2.10}$$

and plugging this definition in Eq. (2.9), the following total current is obtained:

$$I = \frac{2e}{h} \int T(E) \left[f_l(E, V_{SD}) - f_r(E, V_{SD}) \right] dE. \tag{2.11}$$

2.3.1 Spin-Dependent Transport in 1-D Systems

For systems with spin-dependent response, it is useful to define new parameters beyond the transmission and current. Assuming that no spin flip occurs within the considered system, the spin-dependent transmissions $T_{\pm}$ are defined as the transmission through the system of a charge carrier with spins up and down, respectively, and the transmission polarization,

$$P_T = \frac{T_+ - T_-}{T_+ + T_-}. \tag{2.12}$$

The current defined in Eq. (2.11) can be straightforwardly extended to a spin-dependent intensity,

$$I_{\pm} = \frac{e}{h} \int T_{\pm}(E, U_G) \Big[f(E, \mu_S) - f(E, \mu_D) \Big] dE. \tag{2.13}$$

Then, the total current through the device is calculated as

$$I = I_+ + I_-, \tag{2.14}$$

and its polarization, as

$$P = \frac{I_+ - I_-}{I_+ + I_-}. \tag{2.15}$$

2.3.2 Current Density of a Mode Within the Tight Binding Model

In a TB calculation, the usual output is the wave function ψ itself. Therefore, it is necessary to calculate the transmission, i.e., the current density associated to ψ. In a quasi-one-dimensional system this can be easily done by using the continuity equation, which relates the time evolution of the probability density $\rho(x,t) = |\psi(x,t)|^2$ with the divergence of the probability current [31]:

$$\frac{\partial}{\partial t}\rho(x,t) = -\frac{\partial}{\partial x} j(x,t), \tag{2.16}$$

which for a discrete system with period a can be expressed as

$$j(n,t) - j(n+1,t) = a\frac{\partial}{\partial t}\rho(n,t). \tag{2.17}$$

The system needs to be divided in slices in such a way that atoms in the nth slice can only be connected to the atoms in the neighbouring slices $n-1$, $n+1$, that is, the blocks of the Hamiltonian $\mathcal{H}$ fulfil $\mathcal{H}_{m,n} \neq \mathbf{0} \Leftrightarrow |m-n| \leq 1$. The wave function at n is defined as $|\psi_n\rangle = \sum_i c_{n,i} |\phi_{n,i}\rangle$. Then,

$$\begin{aligned}
\frac{\partial}{\partial t}\langle\psi_n|\psi_n\rangle &= \left(\frac{\partial}{\partial t}\langle\psi_n|\right)|\psi_n\rangle + \langle\psi_n|\left(\frac{\partial}{\partial t}|\psi_n\rangle\right) \\
&= \frac{i}{\hbar}\sum_m \left(\langle\psi_m|\mathcal{H}_{m,n}|\psi_n\rangle - \langle\psi_n|\mathcal{H}_{n,m}|\psi_m\rangle\right) \\
&= -\frac{2}{\hbar}\mathrm{Im}\left(\mathbf{c}_{n-1}^{*}\mathcal{H}_{n-1,n}\mathbf{c}_n\right) + \mathrm{Im}\left(\mathbf{c}_{n+1}^{*}\mathcal{H}_{n+1,n}\mathbf{c}_n\right),
\end{aligned} \tag{2.18}$$

with $\mathbf{c}_n \equiv (c_{n,1}, \ldots, c_{n,i}, \ldots)$. In the last step the hermiticity of the Hamiltonian, $\mathcal{H}_{m,n}^{*} = \mathcal{H}_{n,m}$, has been used. By simple comparison of Eqs. (2.17) and (2.18), the following expression is obtained:

$$j(n) = \frac{2a}{\hbar}\mathrm{Im}\left(\mathbf{c}_{n-1}^{*}\mathcal{H}_{n-1,n}\mathbf{c}_n\right). \tag{2.19}$$

For stationary solutions, the current density is constant along the sample, $j(n) = j$.

References

1. P.R. Wallace, The band theory of graphite. Phys. Rev. **71**, 622–634 (1947)
2. A.H. Castro Neto et al., The electronic properties of graphene. Rev. Mod. Phys. **81**, 109–162 (2009)
3. S. Reich et al., Tight-binding description of graphene. Phys. Rev. B **66**, 035412 (2002)
4. G.W. Semenoff, Condensed-matter simulation of a three-dimensional anomaly. Phys. Rev. Lett. **53**, 2449–2452 (1984)
5. O. Klein, Die Re exion von Elektronen an einem Potentialsprung nach der relativistischen Dynamik von Dirac. Z. Phys. A: Hadrons Nucl. **53**, 157–165 (1929)
6. R. Saito et al., Trigonal warping effect of carbon nanotubes. Phys. Rev. B **61**, 2981–2990 (2000)
7. F. Guinea et al., Electronic states and Landau levels in graphene stacks. Phys. Rev. B **73**, 245426 (2006)

8. B. Partoens, F.M. Peeters, From graphene to graphite: electronic structure around the K point. Phys. Rev. B **74**, 075404 (2006)
9. Y. Zhang et al., Direct observation of a widely tunable bandgap in bilayer graphene. Nature **459**, 820–823 (2009)
10. D.C. Elias et al., Control of graphene's properties by reversible hydrogenation: evidence for graphane. Science **323**, 610–613 (2009)
11. R. Balog et al., Bandgap opening in graphene induced by patterned hydrogen adsorption. Nat. Mater. **9**, 315–319 (2010)
12. J.G. Pedersen, T.G. Pedersen, Band gaps in graphene via periodic electrostatic gating. Phys. Rev. B **85**, 235432 (2012)
13. F. Guinea et al., Energy gaps and a zero-field quantum Hall effect in graphene by strain engineering. Nat. Phys. **6**, 30–33 (2010)
14. S.-M. Choi et al., Effects of strain on electronic properties of graphene. Phys. Rev. B **81**, 081407 (2010)
15. M. Kim et al., Fabrication and characterization of large-area, semiconducting nanoperforated graphene materials. Nano Lett. **10**, 1125–1131 (2010)
16. L. Yang et al., Quasiparticle energies and band gaps in graphene nanoribbons. Phys. Rev. Lett. **99**, 186801 (2007)
17. M.Y. Han et al., Energy band-gap engineering of graphene nanoribbons. Phys. Rev. Lett. **98**, 206805 (2007)
18. J. Cai et al., Atomically precise bottom-up fabrication of graphene nanoribbons. Nature **466**, 470–473 (2010)
19. L. Döossel et al., Graphene nanoribbons by chemists: nanometer-sized, soluble, and defect-free. Angew. Chem. Int. Ed. **50**, 2540–2543 (2011)
20. D.V. Kosynkin et al., Longitudinal unzipping of carbon nanotubes to form graphene nanoribbons. Nature **458**, 872–876 (2009)
21. L. Jiao et al., Narrow graphene nanoribbons from carbon nanotubes. Nature **458**, 877–880 (2009)
22. J. Schelter et al., Interplay of the Aharonov-Bohm effect and Klein tunneling in graphene. Phys. Rev. B **81**, 195441 (2010)
23. K. Wakabayashi et al., Electronic transport properties of graphene nanoribbons. New J. Phys. **11**, 095016 (2009)
24. W. Greiner, *Relativistic Quantum Mechanics: Wave Equations* (Springer, Berlin, 2000)
25. P. Koskinen et al., Self-passivating edge reconstructions of graphene. Phys. Rev. Lett. **101**, 115502 (2008)
26. W.Y. Kim, K.S. Kim, Prediction of very large values of magnetoresistance in a graphene nanoribbon device. Nature Nanotechnol. **3**, 408–412 (2008)
27. O.V. Yazyev, M.I. Katsnelson, Magnetic correlations at graphene edges: basis for novel spintronics devices. Phys. Rev. Lett. **100**, 047209 (2008)
28. M. Büttiker et al., Generalized many-channel conductance formula with application to small rings. Phys. Rev. B **31**, 6207–6215 (1985)
29. D. Ferry, S. Goodnick, *Transport in Nanostructures* (Cambridge University Press, Cambridge, 1997)
30. S. Datta, *Electronic Transport in Mesoscopic Systems* (Cambridge University Press, Cambridge, 1997)
31. E.A. de Andrada e Silva, Probability current in the tight-binding model. Am. J. Phys. **60**, 753–754 (1992)

Chapter 3
Graphene Nanoring as a Quantum Interference Device

In graphene, Klein tunnelling manifests itself as the occurrence of perfect transparency of barriers at normal incidence, as predicted by Katsnelson et al. [1] and observed in experiments later [2]. This peculiar tunnelling would lead to undesired charge leakage in graphene-based devices. While in particle physics it is known that there exist relativistic interactions for which Klein tunnelling is absent [3, 4], it seems that they have no counterpart in graphene and confining electrons is a challenging task while being necessary for many applications. Therefore, a significant amount of effort has been focused on graphene-based nanodevices that could enhance carrier confinement, such as *p–n* junctions [5, 6], superlattices [7–9] and FET [10, 11].

Interference effects of coherent electron transport through graphene nanorings open an alternative possibility of controlling quantum transport without relying on potential barriers. Interference effects in graphene subjected to a perpendicular magnetic field, such as current revivals [12] or Aharonov-Bohm conductance oscillations in ring-shaped devices [13–17], have already been studied. In particular, in [15] it was pointed out that these conductance oscillations are robust under the effects of either edge or bulk disorder. Wu et al. [16] investigated quantum transport through a graphene nanoring theoretically and concluded that the device behaves like a resonant tunnelling one, in which the resonance energy can be tuned by varying the size of the device or the external magnetic field. Effects of an electrostatic potential applied to one of the arms on the Aharonov-Bohm magnetoconductance were discussed in [17].

In contrast to previous studies of magnetically induced interference effects [13–16], this chapter reports on a new design of graphene interference device, where electron transport is controlled *without* applying a magnetic field. In the proposed device, charge carrier transport can be tuned instead by applying a side-gate voltage across a graphene nanoring. This side-gate voltage introduces an asymmetry between the arms of the ring, and therefore a change in the relative phase of the electron wave function in the two arms, leading to a constructive or destructive interference at the drain, which results in conductance oscillations and current modulation.

J. Munárriz Arrieta, *Modelling of Plasmonic and Graphene Nanodevices*,
Springer Theses, DOI: 10.1007/978-3-319-07088-9_3,

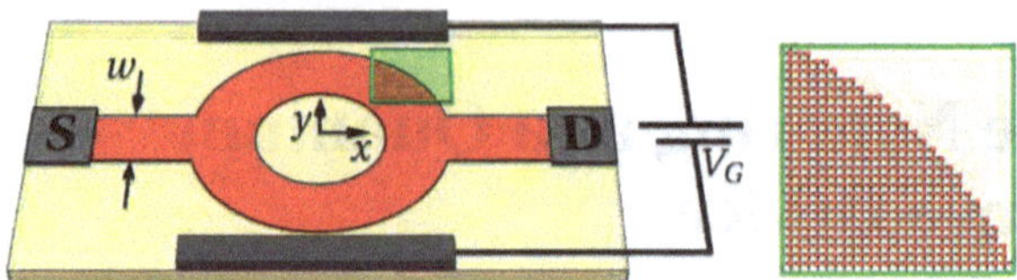

Fig. 3.1 *Left panel* shows the scheme of a *circular ring*. The width w is constant throughout the sample. *Right panel* shows the discretization of the ring inside the region marked with a *green rectangle*

3.1 Background: Interference Effects in Quantum Rings

The first attempts to study interference effects in rings under the effect of external fields were originally performed using 1-D models to characterize the behaviour of charge carriers, if multiple trajectories were allowed [18]. This is the case of the seminal paper by Y. Aharonov and D. Bohm [19]. However, the increase in the computing power led to the possibility of simulating devices without disregarding the effects due to the finite width of the structures.

In mesoscopic systems, a common method to calculate the wave function of the carriers is by discretizing the space using a rectangular mesh. It is straightforward to show that the second derivative present in the Schrödinger equation, upon discretization of the length units Δx, Δy leads to a TB model for a square lattice, with only nearest-neighbour interactions [20], with a hopping parameter

$$t = \frac{\hbar^2}{2m^*a^2}, \tag{3.1}$$

a being the lattice spacing and m^* the effective mass. This type of discretization leads to robust interference effects of the wave function in the ring.

As an example, a circular ring with constant section w was considered (see Fig. 3.1). A discretization scheme was chosen with w split in $N = 25$ sites. An extra side gate was added to the system, so that an asymmetry between the arms could be triggered by applying a voltage V_G. The effect of the side gate is modelled by a profile linear in the y direction and exponentially decaying towards the two leads, with a voltage drop U_G across the sample (more details on the application of the side voltage to the system are given in Sect. 3.2).

The transmission in the low-energy region, with only one propagating mode and voltage drop $U_G = 0$, was calculated and plotted in Fig. 3.2. It is interesting to check that the maxima found in the transmission (central panel) correspond to an increasing number of nodes in the part of the wave function inside each arm of the rings (left and right panels in the same figure).

The application of a side-gate voltage breaks the symmetry between arms. For a constant energy value, maxima and minima alternate with increasing U_G. As expected, maxima are obtained when the difference between the number of nodes in

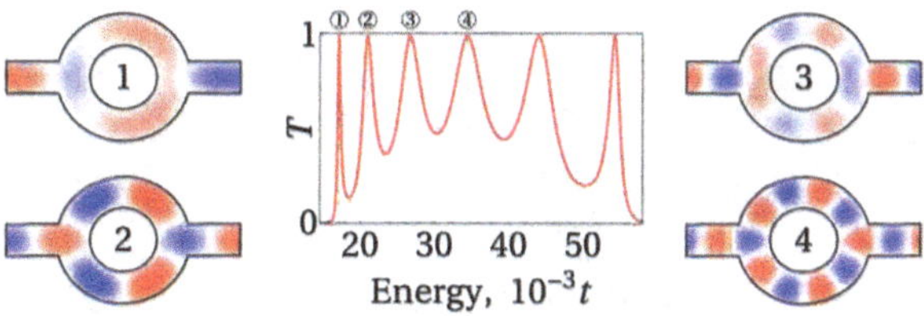

Fig. 3.2 Transmission versus energy in a discretized nanoring ($N = 25$), in the energy region with one mode, without applying a side-gate voltage, i.e., $U_G = 0$. The figures in the *left* and *right* part of the plot show the real part of the propagating wave functions which are solutions of the system for the first maxima, marked with numbers inside *white disks* in the central plot. The number of nodes inside the ring increases with the energy, due to the increase in the k vector of the incoming plane wave

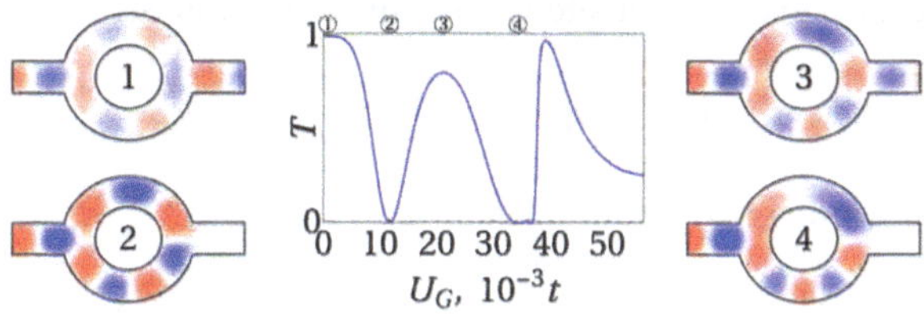

Fig. 3.3 Transmission versus voltage drop due to the side-gate, U_G, in the nanoring. The energy E is fixed in the second maximum of Fig. 3.2, $E = 26.75 \times 10^{-3}$ t. The figures in the *left* and *right* part of the plot show the real part of the propagating wave functions which are solutions of the system for the first maxima and minima, marked with numbers inside *white disks* in the central plot. The number of nodes inside the ring is constant, while the asymmetry between modes in the *upper* and *lower* arms increases

one arm is an even number, odds resulting in transmission minima. This behaviour is exemplified in Fig. 3.3, with the energy being set to the second maximum of Fig. 3.2.

The broad peaks found in the transmission and the smoothness of both $T(E)$ and $T(U_G)$ are a consequence of the regular circular geometry considered. This regularity is partially lost in the edges after the discretization, with steps of different sizes, as shown in the right plot of Fig. 3.1. However, the effect of the irregularities washes out due to the shape of the propagating eigenmode, with a low probability density in the edges. The propagating eigenmodes in an infinite rectangular ribbon with square lattice can be analytically calculated, as it suffices to calculate the set of eigenmodes of the infinite 2-D lattice, with zero probability density in the rows 0 and $N + 1$. Therefore, given the full set of Bloch wave functions

$$\left|\psi^{2D}(\mathbf{k})\right\rangle = \sum_{m,n=-\infty}^{\infty} c_{m,n}(\mathbf{k}) \left|\phi_{m,n}\right\rangle = \sum_{m,n=-\infty}^{\infty} e^{ik_x ma} e^{ik_y na} \left|\phi_{m,n}\right\rangle, \qquad (3.2)$$

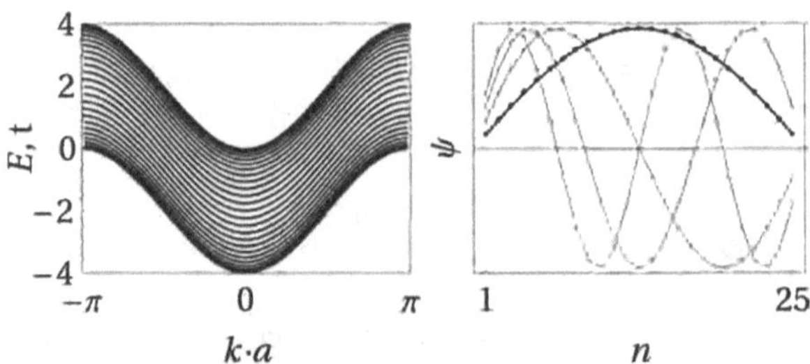

Fig. 3.4 *Left* dispersion relation of an infinite nanoribbon with square lattice and $N = 25$ atoms per column. *Right* transverse section of the first four propagating eigenmodes of the structure. The eigenmode with the smallest energy is highlighted in *black*

$|\phi_{m,n}\rangle$ being the atomic orbital at site (m, n), the condition is

$$c_{i,0} = c_{i,N+1} = 0, \ \forall i \in \mathbb{Z}, \tag{3.3}$$

and the set of allowed eigenvectors shrinks to

$$\left|\psi_i^{1\text{D}}\right\rangle = \sum_{m=-\infty}^{\infty} \sum_{n=1}^{N} e^{imk_x a} \sin(nk_y a), \qquad k_y = \frac{\pi}{N+1} n, \ n \in \mathbb{Z}, \tag{3.4}$$

the dispersion relation being given by

$$E_n(k_x) = -2t \left[\cos(k_x a) + \cos\left(\frac{\pi}{N+1} n\right)\right]. \tag{3.5}$$

These results are plotted in Fig. 3.4. As stated before, for the lowest lying eigenmode, the probability density is mainly in the central part of the ribbon. Therefore, it is to be expected that edge defects in the nanoring have a minor impact on the transmission of the wave function.

This makes an important difference with the case of graphene: as seen in Sect. 2.2, the transport properties in GNRs strongly depend on the edge type. Therefore, for non-regular edges there is strong scattering, which results in very low transmission outside a few sharp resonances. This is exemplified in Fig. 3.5, where the transmission through a circular graphene nanoring of constant width $w = 63\tilde{a}_0 \simeq 15.4\,\text{nm}$ is shown. The wave functions plotted in the left and right panels of the same figure show the presence of strong resonances within the ring.

3.2 System and Modelling

In order to avoid the effect of multiple edge types in the same sample, a device comprising a graphene nanoring with 60° turns attached to semi-infinite leads was proposed. This type of edges has been shown to be experimentally feasible [21].

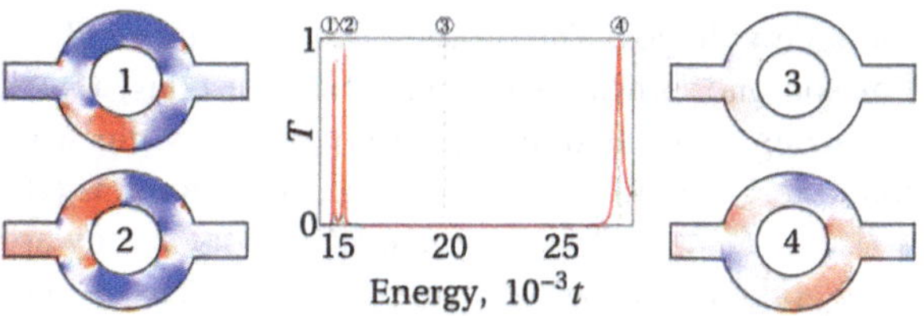

Fig. 3.5 Transmission versus energy in a circular graphene nanoring ($N = 63$), in the energy region with one mode, for $U_G = 0$. The figures in the *left* and *right* part of the plot show the real part of the propagating wave functions which are solutions of the system for the positions marked with numbers inside *white disks* in the central plot. Contrary to the interference patterns found for *square* lattices, the wave functions here show regions with strong resonances

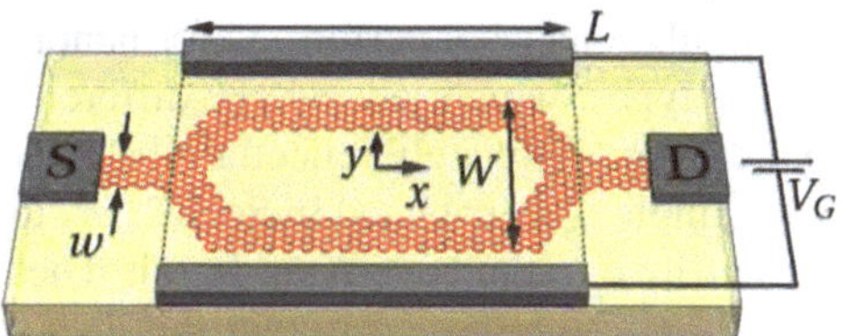

Fig. 3.6 Schematic diagram of the device: a graphene nanoring attached to two leads. The geometry is determined by the parameters L, W, and w. The side-gate voltage V_G is applied across the nanoring as shown in the plot. A back-gate voltage can also be applied to shift the Fermi level

The schematic diagram of the device is shown in Fig. 3.6. The total length of the ring is L, its total width W, while the width of all nanoribbons is w. Note that these dimensions should be large enough to avoid dielectric breakdown at used source-drain and side-gate voltages.

As mentioned in Sect. 2.2, the electronic structure of nanoribbons is very sensitive to the type of edges [22]. Thus, if a nanoribbon has a turn which does not preserve the edge type, a propagating charge carrier would experience strong scattering at the turn due to the electronic structure mismatch, which is disadvantageous for transport. Therefore, the above mentioned design of the quantum nanoring with 60° turns was proposed, which does preserve the edge type and greatly reduces such scattering at the turns.

The device was modelled using Eq. (2.1). If the device is set to operate in an energy range close to the neutrality point for graphene, the interactions can be restricted to nearest neighbours. Therefore,

$$\mathcal{H} = \sum_i \epsilon_i \left|\phi_i\right\rangle \left\langle\phi_i\right| - \sum_{\langle i,j\rangle} t \left(\left|\phi_i\right\rangle \left\langle\phi_j\right| + \left|\phi_j\right\rangle \left\langle\phi_i\right|\right), \tag{3.6}$$

The hopping parameter was set to $t = 2.8\,\text{eV}$ [23], and the spin-related effects, neglected (see, e.g., [24, 25] or [26] for a review). The site energy ϵ_i can depend on the position of the ith atom due to the presence of the source-drain and both back-gate and side-gate voltages. The profile of the electric field can be calculated by solving the Poisson and Schrödinger equations self-consistently. However, for simplicity, the following simplified side-gate potential profile was assumed: it is linear in the y direction ($|y| < W/2$) while in the x direction it is (i) constant within the nanoring area ($|x| \leq L/2$), and (ii) exponentially decaying towards the two leads (for $|x| \geq L/2$). The voltage drop between arms U_G is used as the amplitude, instead of the applied voltage V_G, so that no further assumptions need to be done.

Using the Quantum Transmission Boundary Method (QTBM) [27, 28], adapted for graphene in appendix A, for each energy E and side-gate voltage drop U_G the wave function $|\psi\rangle$ in the whole sample was calculated. Then, using the procedure described in Sect. 2.3, both the transmission coefficient $T(U_G, E)$ and the current-voltage characteristics could be obtained.

Extensive numerical simulations were performed for nanorings with a variety of sizes, geometries and edge types, which led to very different transmission coefficient patterns. These patterns are intimately related to the dispersion relation in the nanoribbons forming the sample. As explained in Sect. 2.2.2, the dispersion relation in a nanoribbon is very sensitive to the edge type. In this Thesis, rings made up of ribbons with edges belonging to the three different families were studied: zGNR, metallic aGNR and semiconducting aGNR. The allowed modes have been plotted in Fig. 3.7. The transverse wavenumber k is measured in units of the inverse lattice spacing a^{-1} along the nanoribbon (i.e., $a = \sqrt{3}\, a_0$ for zigzag edges and $a = 3\, a_0$ for armchair ones, a_0 being the interatomic distance).

As shown below, the gapped aGNR presents more robust and promising transmission patterns for transport control and applications. In all three cases, when the ribbon width is increased, consecutive dispersion branches become closer to each other and the energy region with a small number of propagating eigenmodes shrinks. The focus will be set at the one-mode regime, where interference-related effects are not smoothed out due to the superposition of several modes.

3.3 Results and Discussion

3.3.1 Resonance Transmission Bands in Armchair-Edged Nanorings

The sample addressed in this section has armchair edges and the following geometry: $L = 214\,\text{nm}$, $W = 107\,\text{nm}$ and $w = 15.1\,\text{nm}$ (which corresponds to $N = 62$), comprising $\sim 3 \times 10^5$ atoms. In Fig. 3.8 the transmission coefficient as a function of the Fermi energy and side-gate voltage drop U_G is shown. The transmission pattern consists of a series of very sharp and narrow resonance lines. Outside the resonances the transmission is vanishingly small. To illustrate this more clearly, cross sections of the transmission map were taken along the horizontal red and vertical blue lines

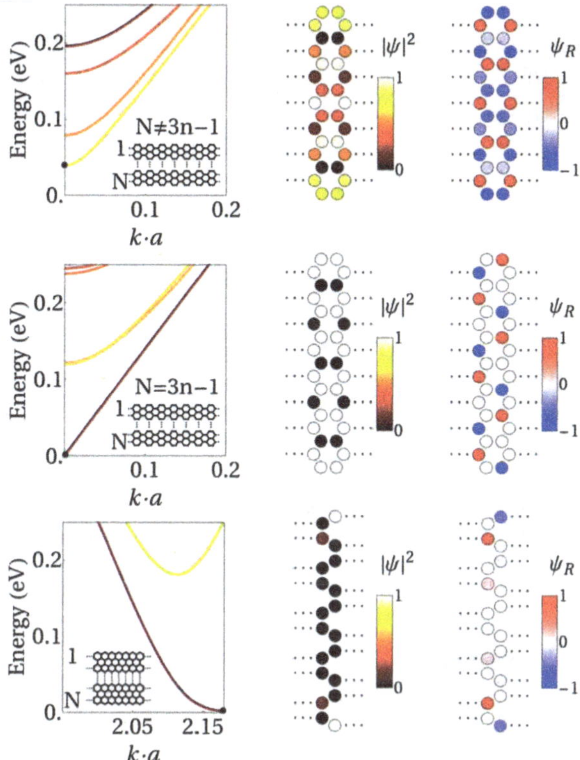

Fig. 3.7 *Left panels* Dispersion relations near the band edges for nanoribbons with armchair edges with $N = 63 \neq 3n - 1$ and $N = 62 = 3n - 1$, $n \in \mathbb{N}$ (*upper* and *central panels*, respectively), and a nanoribbon with zigzag edges and $N = 70$ (*lower panel*). *Middle* and *right panels* probability density and real part of the eigenmodes at the band edges, in one unit cell. For clarity, narrower nanoribbons corresponding to the same families have been used, chosen in positions analogous to the *black points* plotted in the dispersion relation. The edge nature of the lowest mode in zGNR is apparent, as well as the asymmetry of the linear mode for aGNR, observed in the plot of the real part of the wave function

in the upper panel of Fig. 3.8, which corresponds to $U_G = 0$ (lower left panel) and $E = 19.2$ meV respectively (lower right panel).

The wave function of a high transmission state has a huge pile up in the region of the nanoring, which is the typical wave function structure of a resonance state. These results are plotted in Figs. 3.9 and 3.10, for the transmission peaks marked by letters in the lower panels of Fig. 3.8. When no lateral voltage is applied, resonances appear in two different sections, as seen in Fig. 3.9. By increasing the energy of the incoming mode, thus reducing its electronic wavelength, new resonances are triggered, with increasing number of nodes. Interestingly, if the side gate is switched on, the wave function transmits through the ring by using alternately the two resonances states (see Fig. 3.10). It is this swapping between resonances that produces both crossings and

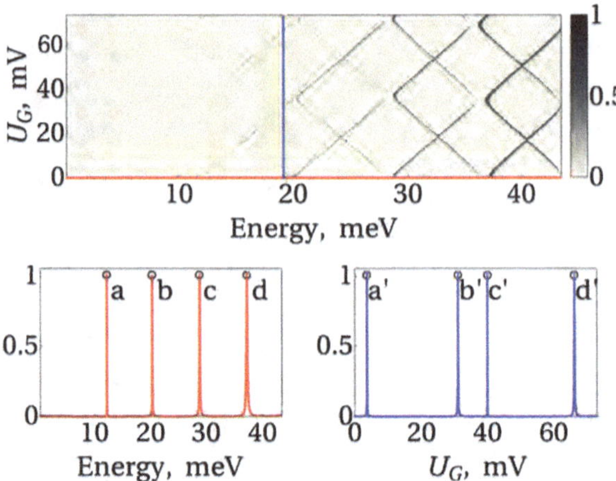

Fig. 3.8 The *upper panel* shows the transmission map for a system with armchair edges and $N = 62$, as a function of the Fermi energy E and the side-gate voltage U_G. *Lower left* and *right panels* show a cross section of the transmission map along the *horizontal red* and *vertical blue lines* in the *upper* plot, corresponding to $U_G = 0$ and $E = 19.2\,\text{meV}$, respectively. Extra points have been calculated along these two lines to enhance the resolution of the peaks

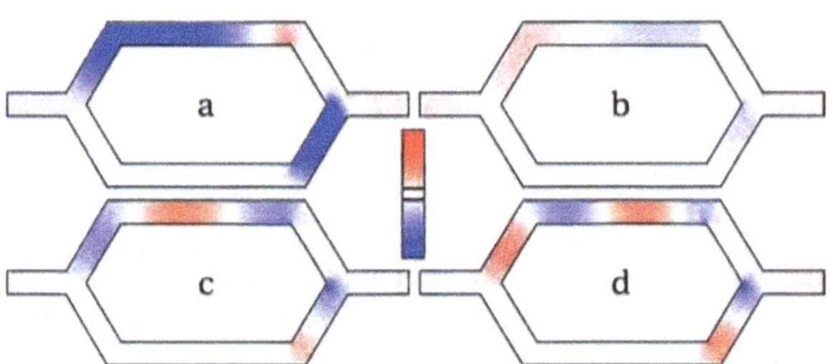

Fig. 3.9 Plots of the real part of the wave functions corresponding to the four peaks of the *lower left panel* in Fig. 3.8, showing strong resonances. The *black lines* in the colour scale correspond to the maximum and minimum real part of the incoming mode

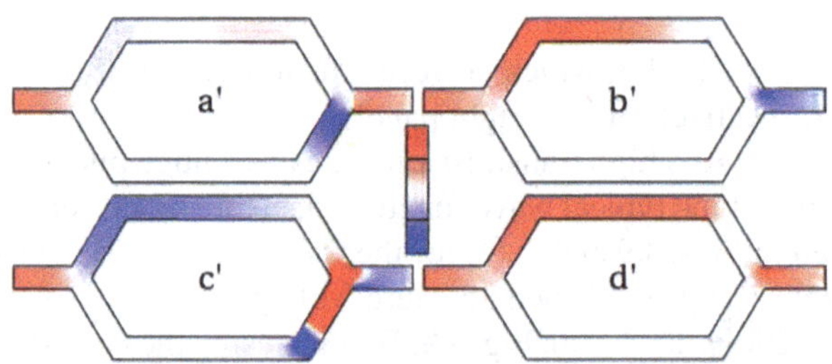

Fig. 3.10 Same as Fig. 3.9, for the four peaks of the *lower right panel* in Fig. 3.8, showing also strong resonances. The transmission through the ring is accomplished by using the two resonant states alternately

anti-crossings when lines with strong transmissions merge (upper panel of Fig. 3.8). The asymmetry in the propagation of the wave function through the two arms of the nanoring when $U_G = 0$ might seem striking at first glance, but it has its origin in the asymmetry of the incoming mode (central panel of Fig. 3.7).

The straight lines observed in the upper panel of Fig. 3.8 can be understood using the following reasoning. The effect of the transverse electric field can be seen as a shift in energy in the upper and lower arms of the nanoring. Therefore, if a side-gate voltage U_G is applied, then the energy shifts in the upper and lower arms can be estimated as $+U_G\,\widetilde{W}/2$ and $-U_G\,\widetilde{W}/2$, respectively. Here $\widetilde{W}$ is some effective width of the nanoring. Therefore, an incoming mode with wave vector $k(E_{\text{in}})$ would propagate through the upper branch with a wave vector $k(E_{\text{in}} + U_G\,\widetilde{W}/2)$ and with $k(E_{\text{in}} - U_G\,\widetilde{W}/2)$ in the lower one. If the resonance condition without transverse field is obtained at E_{res}, then when a small voltage is applied, the resonance splits to two distinct energy points, $E_{\text{res}} \pm U_G\,\widetilde{W}/2$. When the dispersion relation is linear or almost linear, this condition leads to the occurrence of the observed straight lines crossing at $U_G = 0$ in the transmission map. The validity of this simplified picture was checked by calculating the transmission for a system with one single arm, which resulted in a qualitatively similar map but without the lines corresponding to resonances of the removed branch.

The transmission in the vicinity of a resonance can be changed abruptly by a very small variation of the side-gate voltage. This could be very attractive for applications, but this type of device can hardly be practical because the resonances are very narrow and can easily be affected by perturbations, such as disorder. This conjecture was numerically confirmed. Moreover, the mode with linear dispersion is due to very unstable boundary conditions. As stated in Sect. 2.2.2, considering the contribution of the non-nearest neighbours, small as they might be, unavoidably opens a gap, and something similar occurs when edge reconstruction is included. Moreover, it has been shown using ab-initio calculations [29], that intrinsic lattice deformations due to Peierls instabilities also destroy the linear mode.

3.3.2 Interference Transmission Bands in Armchair-Edged Nanorings

In this section, the most promising configuration is studied: a nanoring composed of nanoribbons with armchair edges and $N \neq 3n - 1$ (n being a positive integer) for which the energy spectrum has a gap at the Dirac point. The dispersion relation in this case is parabolic in the vicinity of $k = 0$ (see upper left panel of Fig. 3.7), which makes it similar to a conventional semiconductor. The transmission coefficient presented in Fig. 3.11 manifests two regions: at lower Fermi energies the aforementioned resonant behaviour with very narrow peaks is observed while at higher energies the transmission comprises wider bands, which arise from interference effects. The lower right plot in Fig. 3.11 shows the dependence of the transmission coefficient on

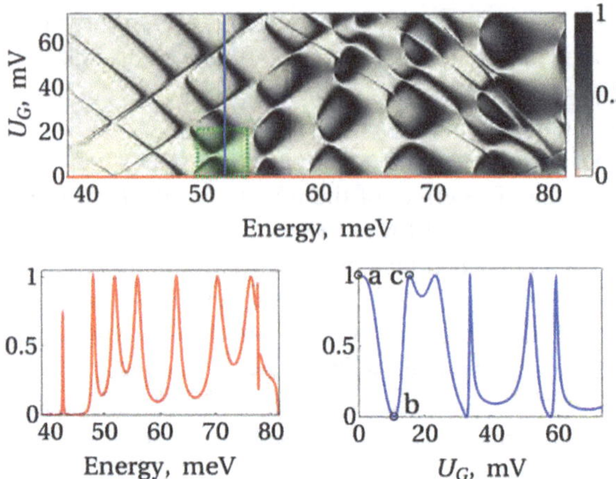

Fig. 3.11 Same as in Fig. 3.8 but for $N = 63$ and $E = 51.9\,\text{meV}$. The *green dashed square* shows the region used to calculate the current-voltage characteristics in Sect. 3.3.4. In the *lower right panel* the first two maxima and the first minimum are highlighted. The corresponding wave functions are plotted in Fig. 3.12

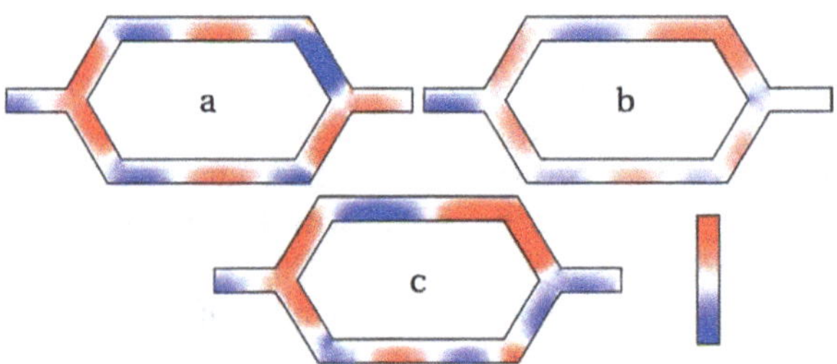

Fig. 3.12 Real part of the envelope wave function corresponding to the three states with different transmission coefficients marked by *open circles* and labelled *a*, *b*, *c* in the *lower right panel* of Fig. 3.11

the gate voltage for a fixed Fermi energy (it corresponds to the cross section of the transmission map along the vertical blue line in the upper panel). Similar plots are obtained for other higher Fermi energies, as long as only one single mode contributes to the transmission.

In order to study the nature of these wider bands the real part of the envelope wave functions are plotted for the three side-gate voltages highlighted in the lower right panel of Fig. 3.11. The system has high transmission for the first and the last energy values (denoted a and c); the corresponding wave functions manifest clear constructive interference patterns at the right lead (upper left and lower panels of Fig. 3.12). Contrary to that, as the upper right panel suggests, the two parts of the low transmission state b are propagating along the two branches of the nanoring in such a way that they arrive to the right extreme of the nanoring out of phase, which

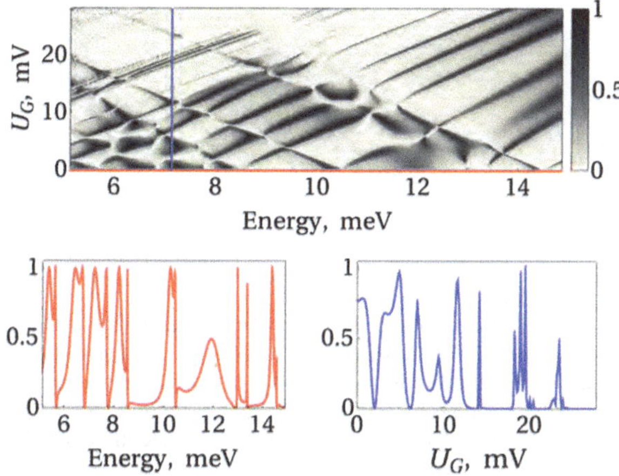

Fig. 3.13 Same as in Fig. 3.8 but for zigzag edges and $N = 70$

gives rise to destructive interference at the drain and practically zero transmission coefficient. Similar patterns were obtained for all other extremes in this higher Fermi energy region.

These interference induced bands are much wider than those having resonance nature and therefore are expected to be more robust and stable with respect to perturbations, such as disorder. Effects of disorder are discussed in the next section, where it will be shown that such a conformation of the nanoribbons comprising the nanoring is the most favourable for applications.

3.3.3 Interference Transmission Bands in Zigzag-Edged Nanorings

For completeness, the sample with zigzag edges is also addressed; the conforming nanoribbons have a width of $w = 15$ nm ($N = 70$). Other geometrical parameters are as follows: $L = 150$ nm and $W = 102$ nm. Such a nanoribbon has a gapless dispersion relation, with low-energy excitations corresponding to high wave numbers k (see left panel of Fig. 3.7). Similar to the previous case of the armchair edged sample, the transmission map also presents interference bands (see Fig. 3.13). However, these bands are considerably narrower than in the case of the armchair edges, which makes them less robust under perturbations and, presumably, less suitable for applications. Moreover, as plotted in the lower panel of Fig. 3.7, the considered mode corresponds to an edge state, and therefore it is prone to localize for small edge distortions.

3.3.4 Current–Voltage Characteristics

In this section the calculations of the current-voltage characteristics of the device with armchair edges and gaped spectrum are presented. The results were obtained

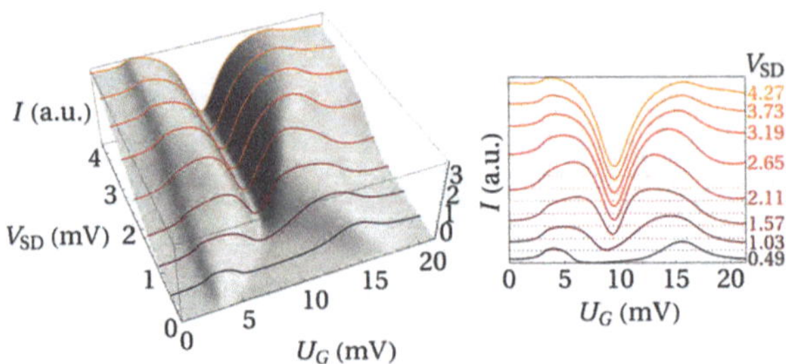

Fig. 3.14 Current-voltage characteristics of the device with armchair edges and the spectrum with a gap. The *left plot* shows $I(U_G, V_{\mathrm{SD}})$, with the Fermi energy initially set to $E_S = 49.6\,\mathrm{meV}$. Some cross-sections with constant V_{SD} are highlighted, and plotted separately in the *right panel*. For clarity, this *lines* have been shifted, the new *bottom lines* being plotted using the same colours. The figures show that intensity can be very efficiently controlled by the side gate

following the guidelines of Sect. 2.3, with the temperature T set to 4 K. The calculation was restricted to the region of the SD and gate voltages marked with the dashed green rectangle in Fig. 3.11. The Fermi energy of both contacts is set to the working point by the back-gate voltage: to the energy E_S (corresponding to the left edge of the rectangle) and then the SD voltage is changed within the selected window.

The corresponding I–V characteristic is presented in Fig. 3.14. The left panel shows the complete $I(U_G, V_{\mathrm{SD}})$ surface. As can be seen from the figure, the side-gate voltage can be used to control the current through the device. The right panel of the figure shows the dependence of the current on the side-gate voltage U_G for several fixed values of SD voltage (specified in the right side of the figure, using the same colour). The on/off ratio of this quantum interference transistor can be as high as about 10. It should be stressed that the current-voltage dependencies typical for a traditional FET are monotonous functions of the gate voltage. However, the proposed device manifests more interesting gate voltage dependence. In particular, the I–V curves have negative differential resistance parts, which can be very useful for applications. Another underlying difference between a traditional FET and the proposed device is the principle of operation. In the latter case it is based upon an essentially quantum mechanical effect: the interference between the two parts of the wave function propagating along the two arms of the nanoring.

Several designs of nanodevices based on single organic molecules exploiting various quantum mechanical effects have already been put forward (see, e.g., [30, 31] and references therein). In particular, Stafford et al. studied quantum interference effects in aromatic molecules as a method to modulate the current flow [30]. More complex organic molecules, such as the DNA, have been proposed to design FETs and more sophisticated devices [31]. However, single-molecule electronics often requires almost atomic level control of contacts; it can be affected greatly by vibrations and could be subject to structural instabilities under required voltages. Fabrication of graphene nanorings seems to be more feasible (at least nowadays), and they can also sustain higher voltages and currents, which is advantageous for applications.

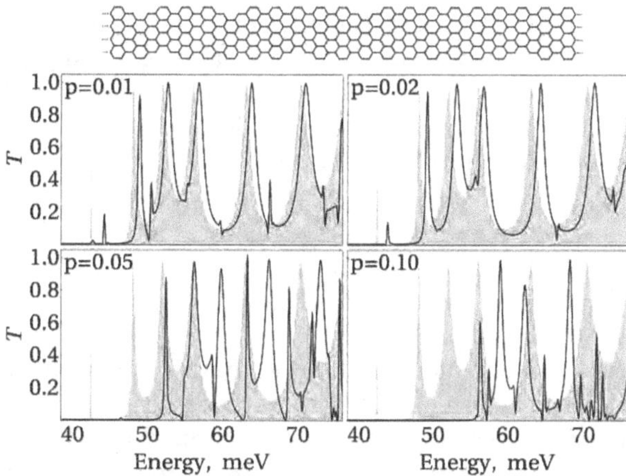

Fig. 3.15 Effect of disorder on the transmission. *Black lines* are used to plot the transmission coefficients versus the mode energy for four different disorder realizations, as compared with the ordered sample—*light grey* regions. Couples of atoms from the edges have been removed with probability $p = 0.01, 0.02, 0.05, 0.10$, as illustrated in the *upper panel*

3.4 Effects of the Edge Disorder

The effects of edge disorder on the transport properties of the proposed device are addressed in this section, using the sample with armchair edges and $N = 63$ to study these effects. To do so, pairs of carbon atoms are removed from the edges with some given probability p (see upper panel of Fig. 3.15). By removing pairs rather than individual atoms the presence of dangling atoms in the sample is avoided, which allows to neglect complicated edge reconstruction effects [32].

The transmission coefficient calculated for particular realizations of disorder for $p = 0.01, 0.02, 0.05, 0.10$ and zero side-gate voltage is presented in Fig. 3.15, with black lines representing the transmission coefficient of the disordered sample. For comparison purposes, grey regions are plotted, representing the transmission in the ordered sample. Two important trends are to be mentioned. First, increasing disorder causes the transmission bands to be shifted towards higher energies with respect to their positions in the regular sample. This is due to the fact that the removal of atoms from the edges of the nanoribbons makes them effectively narrower, which leads to higher quantization energy in the lateral direction. Second, anti-resonances in the bands appear in all the disorder configurations. Several mechanisms could be responsible for this anti-resonances. A possible explanation is the presence of edge states localized by the disorder. However, this possibility was ruled out by performing simulations with the same type of disorder in nanoribbons. These simulations did not show any trace of anti-resonances. Another mechanism is the symmetry breaking between both arms, which leads to a breaking of the degeneracy of the resonances corresponding to both arms. This possibility is further explored in Sect. 4.3.3.

The most important finding is that transmission bands are not destroyed by moderate disorder, and therefore such a device (with armchair edges and gaped spectrum) is robust under perturbations. On the other hand, nanorings having different configurations (zigzag edges or armchair edges with linear dispersion) were found to be affected by the disorder to a much larger extent.

3.5 Summary

In summary, a new quantum interference device based on a graphene nanoring with 60° turns was proposed and studied. Transport properties of the device were found to be very sensitive to the type of edges (zigzag or armchair). The ring comprised of nanoribbons with armchair edges and parabolic dispersion relation with a gap proved to be the most advantageous for electronic transport because the transmission pattern presents wide bands of high transitivity in this case. It was shown that the current flow through the device can be controlled by the side-gate voltage. Such a voltage changes the relative phase of the electron wave function in the two arms of the ring resulting in constructive or destructive interferences at the drain. Consequently, the current flow can be modulated efficiently without applying a magnetic field, so the device operates as a quantum interference effect transistor, which was shown to be robust under moderate edge disorder.

It should be pointed out that the proposed device must be operating in the single mode regime in order to use the interference effects in their most pure form. When the second mode comes into play the interference bands smear out and the current control is expected to be less efficient. In this regard, the dispersion relations of the nanoribbons constituting the device provide an important starting point because they define the appropriate energy window where one single mode is contributing to the transport. For the considered nanoribbon width of about 15 nm such a window is on the order of 40 meV (see upper left panel of Fig. 3.7). As the width w increases the window shrinks while its lower edge approaches the Dirac point. On the other hand, electronic transport through wider nanoribbons is less affected by the edge disorder. These considerations should be taken into account when designing and fabricating the real world device.

References

1. M.I. Katsnelson et al., Chiral tunnelling and the Klein paradox in graphene. Nat. Phys. **2**, 620–625 (2006)
2. N. Stander et al., Evidence for Klein tunneling in graphene p-n junctions. Phys. Rev. Lett. **102**, 026807 (2009)
3. S. Ru-keng, Z. Yuhong, Exact solutions of the Dirac equation with a linear scalar confining potential in a uniform electric field. J. Phys. A: Math. Gen. **17**, 851 (1984)

4. F. Domínguez-Adame, A relativistic interaction without Klein paradox. Phys. Lett. A **162**, 18–20 (1992)
5. J.R. Williams et al., Quantum Hall effect in a gate-controlled p-n junction of graphene. Science **317**, 638–641 (2007)
6. H.-Y. Chiu et al., Controllable p-n junction formation in monolayer graphene using electrostatic substrate engineering. Nano Lett. **10**, 4634–4639 (2010)
7. C. Bai, X. Zhang, Klein paradox and resonant tunneling in a graphene superlattice. Phys. Rev. B **76**, 075430 (2007)
8. H. Sevinçli et al., Superlattice structures of graphene-based armchair nanoribbons. Phys. Rev. B **78**, 245402 (2008)
9. M. Yang et al., Two-dimensional graphene superlattice made with partial hydrogenation. Appl. Phys. Lett. **96**, 193115 (2010)
10. X. Wang et al., Room-temperature all-semiconducting sub-10-nm graphene nanoribbon field-effect transistors. Phys. Rev. Lett. **100**, 206803 (2008)
11. Y. Lu et al., High-on/off-ratio graphene nanoconstriction field-effect transistor. Small **6**, 2748–2754 (2010)
12. E. Romera, F. de los Santos, Revivals, classical periodicity, and zitterbewegung of electron currents in monolayer graphene. Phys. Rev. B **80**, 165416 (2009)
13. S. Russo et al., Observation of Aharonov-Bohm conductance oscillations in a graphene ring. Phys. Rev. B **77**, 085413 (2008)
14. M. Zarenia et al., Simplified model for the energy levels of quantum rings in single layer and bilayer graphene. Phys. Rev. B **81**, 045431 (2010)
15. J. Wurm et al., Graphene rings in magnetic fields: Aharonov-Bohm effect and valley splitting. Semicond. Sci. Technol. **25**, 034003 (2010)
16. Z. Wu et al., Quantum tunneling through graphene nanorings. Nanotechnology **21**, 185201 (2010)
17. J. Schelter et al., Interplay of the Aharonov-Bohm effect and Klein tunneling in graphene. Phys. Rev. B **81**, 195441 (2010)
18. J.-B. Xia, Quantum waveguide theory for mesoscopic structures. Phys. Rev. B **45**, 3593–3599 (1992)
19. Y. Aharonov, D. Bohm, Significance of electromagnetic potentials in the quantum theory. Phys. Rev. **115**, 485–491 (1959)
20. S. Datta, *Electronic Transport in Mesoscopic Systems* (Cambridge University Press, New York, 1997)
21. X. Li et al., Chemically derived, ultrasmooth graphene nanoribbon semiconductors. Science **319**, 1229–1232 (2008)
22. Y.-W. Son et al., Energy gaps in graphene nanoribbons. Phys. Rev. Lett. **97**, 216803 (2006)
23. A.H. Castro Neto et al., The electronic properties of graphene. Rev. Mod. Phys. **81**, 109–162 (2009)
24. H. Şahin, R.T. Senger, First-principles calculations of spin-dependent conductance of graphene akes. Phys. Rev. B **78**, 205423 (2008)
25. H. Şahin, S. Senger, R.T. Ciraci, Spintronic properties of zigzag-edged triangular graphene akes. J. Appl. Phys. **108**, 074301 (2010)
26. O.V. Yazyev, Emergence of magnetism in graphene materials and nanostructures. Rep. Prog. Phys. **73**, 056501 (2010)
27. C.S. Lent, D.J. Kirkner, The quantum transmitting boundary method. J. Appl. Phys. **67**, 6353–6359 (1990)
28. D.Z.Y. Ting et al., Multiband treatment of quantum transport in interband tunnel devices. Phys. Rev. B **45**, 3583–3592 (1992)
29. S.-H. Lee et al., Band gap opening by two-dimensional manifestation of peierls instability in graphene. ACS Nano **5**, 2964–2969 (2011)
30. C.A. Stafford et al., The quantum interference effect transistor. Nanotechnology **18**, 424014 (2007)
31. A.V. Malyshev, DNA double helices for single molecule Electronics. Phys. Rev. Lett. **98**, 096801 (2007)
32. Ç.Ö. Girit et al., Graphene at the edge: stability and dynamics. Science **323**, 1705–1708 (2009)

Chapter 4
Graphene Nanoring as a Source of Spin-Polarized Electrons

As outlined in the introduction to this Thesis, graphene is a material with a combination of many remarkable properties, in particular, large electron mobility and long spin-coherence lengths (up to several microns) [1–3]. These features have spurred the interest in graphene as a material for spintronic devices which exploit both the charge and the spin degrees of freedom as the basis of their operation. Geim and co-workers used soft magnetic NiFe electrodes to inject polarized electrons into graphene and found spin valve effects [4]. Later, Cho et al. performed four-probe spin-valve experiments on graphene contacted by ferromagnetic Permalloy electrodes [5]. The different geometries of the contacts resulted in different coercive fields. Then, they could perform measurements with both parallel and antiparallel configurations. The observation of sign switchings in the nonlocal resistance indicates the presence of a spin current between injector and detector. The drift of electron spins under an applied dc electric field in spin valves in a field-effect transport geometry at room temperature was studied in [6]. These experiments were found to be in quantitative agreement with a drift-diffusion model of spin transport. More recently, Dedkov et al. proposed that the Fe_3O_4 /graphene/Ni trilayer can also be used as a spin-filtering device, where the half-metallic magnetite film was used as a detector of spin-polarized electrons [7].

With the development of the nanoscale technology of graphene, a number of nanodevices have been proposed to explore novel spin-dependent transport phenomena. Spin filter effects in graphene nanoribbons with zigzag edges were investigated theoretically by Niu and Xing using a non-equilibrium Green's function method [8]. They found a fully polarized spin current through ferromagnetic graphene/normal graphene junctions, whose spin polarization could be manipulated by adjusting the chemical potential of the leads. Ezawa investigated similar effects in a system made of graphene nanodisks and leads, where the magnetic moment of the nanodisk can be controlled by the spin current [9]. Guimarães et al. studied spin diffusion in metallic graphene nanoribbons with a strip of magnetic atoms substituting carbon ones in the honeycomb lattice [10]. They found that the system behaves as a spin-pumping transistor without net charge current. More recently, Zhai and Yang have shown that the combined effects of strained and ferromagnetic graphene junctions can be used to fabricate a strain-tunable spin filter [11]. All these findings open the possibility of

J. Munárriz Arrieta, *Modelling of Plasmonic and Graphene Nanodevices*,
Springer Theses, DOI: 10.1007/978-3-319-07088-9_4,

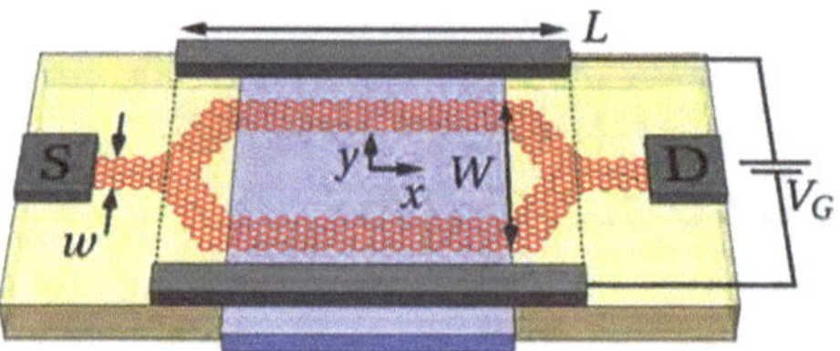

Fig. 4.1 Schematic view of the graphene nanoring fabricated above a ferromagnetic strip (shown as the *blue bar* in the figure). Source and drain terminals are denoted as S and D respectively. Dimensions w, W and L are given in the text

designing spintronic devices based on graphene nanostructures for memory storage and spin diodes. A more complete and detailed review on the electronic and spin properties of mesoscopic graphene structures and control of the spin can be found in [12] and references therein.

In this chapter, the QID studied in Chap. 3 is modified, so that it can be used as an efficient spin filter device. The latter can be achieved by depositing a ferromagnetic insulator below (or above) the nanoring. This deposition has been done experimentally for EuO above graphene [13], and the results suggest that, to a large extent, the structure of graphene remains unaffected. In this case, the combination of the exchange splitting due to the interaction of the electron spin with the magnetic ions and the effect of the side-gate voltage can result in a controllable spin-polarized electric current.

4.1 System and Modelling

The QID to be used is based in the one from Chap. 3. It consists of a graphene nanoring with 60° bends attached to two graphene nanoribbons. The latter are connected to source and drain terminals, as shown in Fig. 4.1. The width of all nanoribbons is w. Two lateral electrodes allow a side-gate voltage to be applied. The total length of the ring is L while the total width is W.

The sample used throughout this work consists of nanoribbons with armchair edges N cells wide, $N \neq 3n - 1$. As shown in Fig. 3.7, this type of nanoribbons presents a gap, with the band edges centred around $k = 0$. In Chap. 3, this type of nanoribbon was found to be the most advantageous one for electronic transport because the transmission spectrum of such a system presents wide bands of high transmission probability.

As a reminder, it should be pointed out that the applied side-gate voltage results in different energy shifts of the electronic states in the two arms of the ring. Thus, a charge carrier injected from the source nanoribbon couples to different modes of the two arms. These two modes can interfere constructively or destructively at the drain, giving rise to conductance and current modulation depending on the side-gate

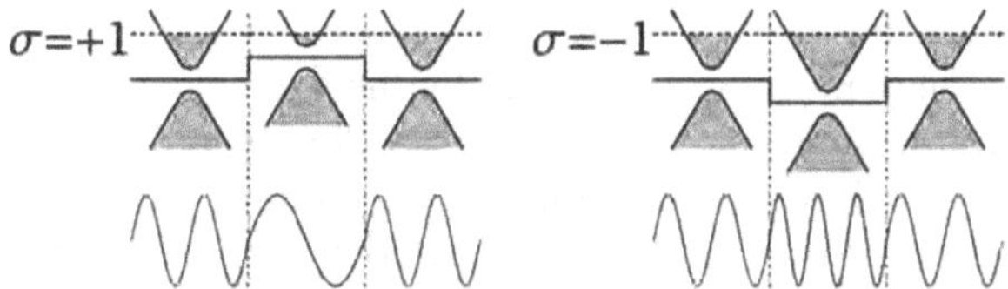

Fig. 4.2 Schematic representation of the potential profile along one of the arms of the ring for spin up (*left*) and spin down (*right*) states. The potential (*solid line*) is spin dependent in the middle section of the ring due to the exchange splitting. Dispersion relations of a nanoribbon with armchair edges and band fillings up to the Fermi energy (*dashed line*) are also shown schematically. The exchange interaction shifts locally the wave vector of the mode towards lower (*higher*) k-values, thus resulting in smaller (*bigger*) wavelengths

voltage. This form of current control relies on interference effects which depend quantitatively on details of device geometry, material parameters, perturbations, such as the disorder, etc. However, the underlying principle of operation is very basic and, as long as propagating modes exist in the two arms of the ring, the control is expected to be feasible. Therefore, simple models are used which grasp the main features of the different system components.

In the present design the nanoring is fabricated above a strip of a ferromagnetic insulator, such as EuO. The exchange interaction between Eu^{2+} ions and charge carriers can be described as an effective Zeeman splitting of the spin sublevels [14]. This creates spin-dependent potential profiles along the arms of the ring (see Fig. 4.2). An injected electron couples to different modes of the arms depending on its spin. Therefore, for some side-gate voltages, the interference governing the conductance can be constructive for spin-up and destructive for spin-down states (or vice versa), resulting in a spin-polarized total current.

The device was modelled using Eq. (2.1) and set to operate at a Fermi energy close to the one for intrinsic graphene. Therefore, the interatomic interactions can be limited to $l = 1$, that is, restricted to nearest neighbours:

$$\mathcal{H} = \sum_i \epsilon_i \left|i\right\rangle \left\langle i\right| - \sum_{\langle i,j\rangle} t \left|i\right\rangle \left\langle j\right| + \sigma\, \Delta_{\text{ex}} \sum_{i \in \mathcal{L}} \left|i\right\rangle \left\langle i\right| \,, \tag{4.1}$$

where the site energy ϵ_i depends on the position of the ith carbon atom, in particular, due to the side-gate voltage. The same simplified side-gate potential profile as in Chap. 3 is used: if the origin of the coordinate system is in the geometrical centre of the ring (as indicated in Fig. 4.1), this potential is linear in the y direction for $|y| < W/2$ while in the x direction it is (i) constant within the nanoring area ($|x| \leq L/2$) and (ii) decays exponentially towards the two leads (for $|x| \geq L/2$). The contacts are assumed to be far enough from the ring, so that the side-gate potential can be safely set to zero at the leads. The total potential drop between the outer edges of the two arms (separated by the distance W) is denoted as U_G and is referred to as the side-gate voltage from now on.

The effect of the ferromagnet is taken into account in the third term of Eq. (4.1). It affects the site energies ϵ_i, shifting them by the amount $\sigma\Delta_{\text{ex}}$, where Δ_{ex} is the exchange splitting amplitude and $\sigma = +1$ ($\sigma = -1$) for spin up (spin down) states. The value $\Delta_{\text{ex}} = 3\,\text{meV}$ was used, which is of the order of the values known from the literature, 3–10 meV [14–16]. The characteristic length scale of the exchange interaction is of the order of one monolayer thickness [14]. Then, only the sites which are in touch with the ferromagnetic strip are affected by the interaction. The set of such sites belonging to the longitudinal sections of the arms (see Fig. 4.1) is denoted as $\mathcal{L}$ in Eq. (4.1). Due to such local splitting, a spin up (down) electron propagating along one of the arms is subjected to the potential with a rectangular barrier (well). Such potential profile is shown schematically in Fig. 4.2 for the case of zero side-gate voltage. As the size of the sample is much smaller than the spin-coherence length, no spin-flip is considered, resulting in independent propagation channels for each spin.

The graphene lattice is known to undergo reconstruction at nanoribbon edges, which affects the corresponding site energies ϵ_i and hoppings [17]. These effects are not expected to play a crucial role for transport properties of realistic disordered samples, and therefore are neglected and an undistorted honeycomb lattice considered with the usual nearest neighbour coupling $t = 2.8\,\text{eV}$ [18].

The wave functions and spin-dependent transmission coefficients $T_{\pm}$ for spin up (+) and spin down (−) electrons were calculated using the same numerical techniques as in the previous chapter (see appendix A for details). These coefficients depend on the energy of the injected carrier E and the side-gate voltage U_G. Using them, the degree of transmission polarization P_T defined in Eq. (2.12), as well as the spin-polarized currents $I_{\pm}$ Eq. (2.11), can be calculated. The source-drain voltage V_{SD} is assumed to drop in the leads, which agrees with recent experimental results [19], and therefore the effect of this voltage drop is modelled as a shift of the Fermi level of the source, μ_S, with respect to that of the drain, μ_D.

In this work, the set of parameters was chosen so that the system operated in the one-mode regime. This implies that the lateral quantization energy (due to the finite nanoribbon width w) is much larger than the source-drain potential drop eV_{SD} and the thermal energy $k_{\text{B}}T$. Further constraints for the latter two parameters are discussed below.

4.2 Results and Discussion

The geometry of the considered sample is the same as in Sect. 3.3.2, i.e., a nanoring made of nanoribbons of width $w = 15.5\,\text{nm}$ with armchair edges, with full ring length set to $L = 179\,\text{nm}$ and width $W = 108\,\text{nm}$ (see Fig. 4.1 for the schematics of the device).

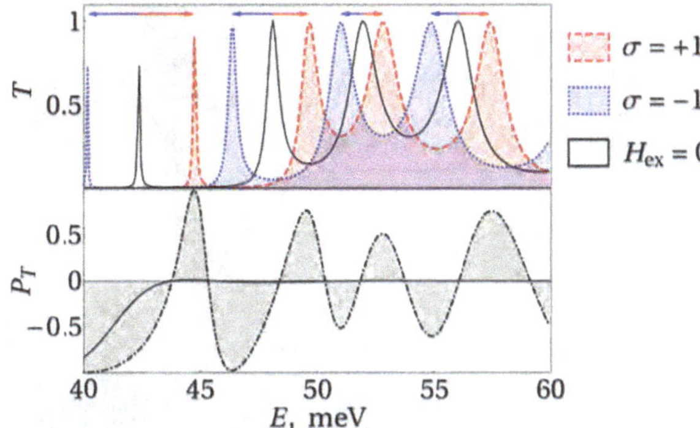

Fig. 4.3 *Upper panel* shows the transmission coefficient calculated without the exchange splitting (*solid black line*) and with it (*dashed red* and *dotted blue lines* correspond to T_+ and T_-, respectively). *Lower panel* shows the transmission polarization P_T of the nanoring (*dot-dashed line*) and that of a single nanoribbon (*solid*). All quantities are calculated at $U_G = 0$

4.2.1 Spin-Dependent Transmission Without Side-Gate Voltage

First, the device properties at zero side-gate voltage were addressed. Transmission coefficients as functions of the carrier energy E are shown in the upper panel of Fig. 4.3. If the substrate is not ferromagnetic there is no exchange splitting ($\Delta_{ex} = 0$) and the spin does not play any role, so $T_\pm$ are degenerate (see the black curve in the figure, the same as the lower left panel of Fig. 3.11). In this reference case the transmission is characterized by a series of peaks which become wider as the carrier energy increases. The interaction with the ferromagnet shifts these features towards lower or upper energies depending on the sign of the carrier spin (see the red dashed curve giving T_+ and the blue dotted one giving T_-). However, the transmission spectrum remains qualitatively the same except for the energy shift. Peak shifts are shown using arrows above the curves in the upper panel.

Lower panel of Fig. 4.3 shows the transmission polarization, as defined by Eq. (2.12), demonstrating that the transmission is highly polarized within some energy ranges. The latter can give rise to the spin-polarized electric current. The polarization degree is higher at lower energies. However, the transmission coefficient in this energy range comprises narrow resonance peaks which can easily be destroyed by perturbations such as disorder, as shown below. The higher energy range with its wider transmission bands, which was proved in Sect. 3.4 to be more robust under the effects of disorder, is more promising for applications. For comparison, the transmission polarization for a single nanoribbon of the same width w and length L as those of the quantum ring was also calculated (see the solid black line in the lower panel of the figure). As expected, the interference effects are absent and the transmission polarization disappears quickly as the carrier energy increases.

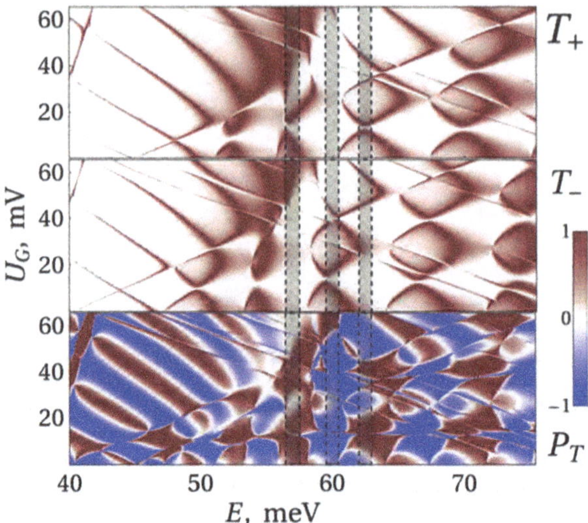

Fig. 4.4 Transmission coefficients $T_\pm$ and degree of the transmission polarization P_T as functions of the carrier energy E and the side-gate voltage U_G

4.2.2 Control of the Current Polarization via the Side-Gate Voltage

Next, the effect of the side-gate voltage on the polarization of the electric current is studied. To this end, the transmission coefficient maps were calculated for spin up and spin down electrons. These results are shown in the upper and middle panels of Fig. 4.4. Using them, the transmission polarization degree P_T was calculated, and presented in the lower panel of the figure, which demonstrates that the polarization can be controlled by the side-gate voltage. Hereafter, the focus is set on the higher energy range ($E > 50$ meV in the considered case) where wider transmission bands have the interference nature. As shown in that figure, the sign of the polarization is almost independent of the side-gate voltage for some energies (see the leftmost and the middle shadowed strips), while for others both the polarization degree and its sign can be changed by the side-gate (see, e.g., the rightmost shadowed strip), which opens the possibility to control the polarization of the current by the electrostatic gate.

To demonstrate this possibility, the total current I, together with its polarization degree P, were calculated, using Eqs. (2.14) and (2.15). The Fermi energies of both the source and the drain were set to some value by a back-gate voltage and then, one of these levels shifted with respect to the other by the source-drain voltage V_{SD}. Upper and lower limits of the three shadowed strips in Fig. 4.4 give the source and drain Fermi energies μ_S and μ_D, respectively, which were used to calculate electric currents using Eq. (2.13) for $V_{\text{SD}} = 1$ mV and $T = 4$ K. This choice of values was suggested by the following reasoning. At $T = 0$ the Fermi distributions are step functions, so only the energy range between μ_S and μ_D contributes to the current [see the integrand in Eq. (2.13)]. If this range is greater than the typical energy separation between the transmission bands of T_+ and T_- (see Fig. 4.4), then both polarizations would contribute to the total current to a similar extent and its

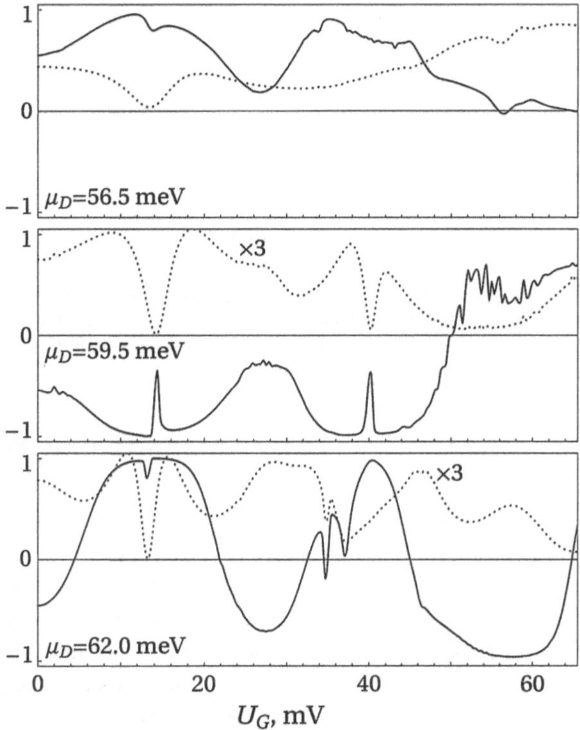

Fig. 4.5 Total electric current I through the device (*dotted lines*) and its polarization degree (*solid lines*) as functions of the side-gate voltage U_G. I is given in units of the current obtained for the case of a perfectly transmitting sample $[T_{\pm}(E, U_G) \equiv 1]$. *Upper, middle* and *lower* panels correspond to the leftmost, middle and rightmost shadowed strips in Fig. 4.4 (see text for details)

polarization would be reduced. Therefore, to observe substantial current polarization, eV_{SD} should be smaller than the typical energy scale ΔE in the transmission maps. Figure 4.4 suggests that this scale is of the order of 5 meV, which justifies our choice of $V_{\mathrm{SD}} = 1\,\mathrm{mV}$. Similar arguments apply to the temperature which smooths the Fermi distribution (a step function at $T = 0$). This increases the range of energies contributing to the current and reduces its polarization. Thus, the temperature should be smaller than $\Delta E/k_{\mathrm{B}} \approx 60\,\mathrm{K}$. Note that this ranges of parameters scale with the exchange interaction. Thus, they would be broadened if the exchange interaction is found to be greater than our conservative estimation.

The total current I (dotted lines) and its polarization degree P (solid lines) are shown in Fig. 4.5. Upper, middle and lower panels correspond to the leftmost, middle and rightmost shadowed strips in Fig. 4.4, respectively. As expected, the electric current can be highly polarized. For some combinations of the source and drain Fermi energies, the sign of the current polarization was found to remain the same within wide ranges of the side-gate voltage (see the upper and the middle panels of Fig. 4.5). Nevertheless, as can be seen from the lower panel, the Fermi energies at the source and drain can be adjusted in such a way that the current polarization can be changed in almost its entire possible range $[-1, 1]$ by the side-gate voltage, suggesting that the proposed device operates as a controllable source of spin-polarized electrons.

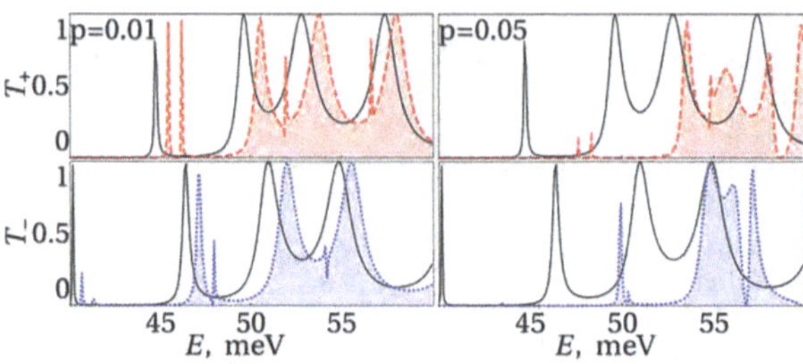

Fig. 4.6 Transmission coefficients $T_{\pm}$ of the regular sample (*solid black lines*) and those of a disordered one (*curves with filling*), after removing couples of atoms from the edges with probabilities $p = 0.01$ (*left*) and $p = 0.05$ (*right*). The side-gate voltage U_G is set to zero

4.3 Effect of Edge Disorder and Geometric Imperfections

In the aforementioned calculations, an ideal symmetric nanoring was considered, while different imperfections or perturbations, in particular, the disorder, could affect the electric current and its polarization. There are various possible sources of disorder, for example, charged impurities in the substrate or defects of the device fabrication, such as imperfections of the device edges. While the former would provide some additional smooth electrostatic potential and can hardly deteriorate the transmission through the device to a large extent, the impact of the latter on the transport properties is probably stronger, especially for small devices.

4.3.1 Edge Disorder

In order to estimate a possible impact of the edge disorder on the transport properties, pairs of carbon atoms from the nanoring edges were removed with some given probability p, following the recipe given in Sect. 3.4. Transmission coefficients $T_{\pm}$ calculated for two particular realizations of such a disorder ($p = 0.01, 0.05$) and zero side-gate voltage are presented in Fig. 4.6, respectively. Solid black lines show the transmission coefficients of the reference ordered sample while lines with filling show those of a disordered one. The results are similar to those obtained in the previous chapter, i.e., in the disordered sample all transmission bands are shifted to higher energies with respect to their positions in the regular one, as the ribbons are made effectively narrower. This leads to higher quantization energy in the lateral direction that manifests itself in the observed shifts. As expected, narrow resonance peaks in the lower energy range ($E < 47$ meV) are smashed as the disorder increases, while wider interference-related bands at higher energies are not destroyed by the disorder. These bands are affected by the disorder to a comparable degree for both spin up and spin down electrons, which suggests that a moderate disorder would not deteriorate polarization properties of the spin filter to a large extent.

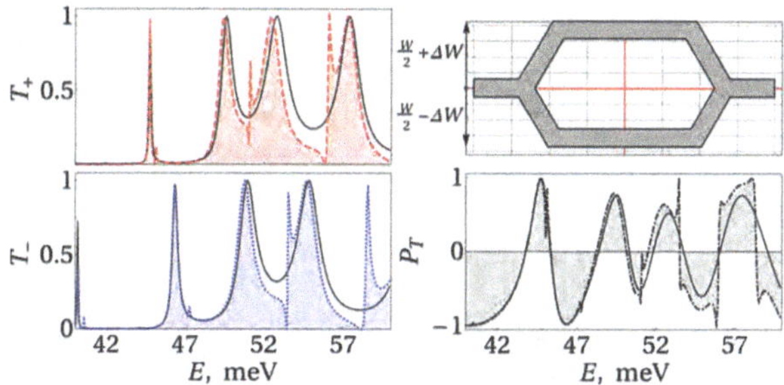

Fig. 4.7 Transmission coefficients $T_{\pm}$ (*upper left* and *lower left panels*) and transmission polarization P_T (*lower right panel*) for an asymmetric nanoring. The ring was engineered by setting the total width of every arm to $W \pm \Delta W$ (see *upper right panel*), while keeping fixed the parameters L, W, w and the 60° turns. In all the graphs, the output of the asymmetric structure is plotted using *filled lines*, while those corresponding to the symmetric one use *solid black lines*

4.3.2 Geometric Imperfections

Finally, possible effects of a different kind of imperfections in the device fabrication on its performance were addressed. The analysis is performed by dividing these imperfections in different fundamental types. One of them could be a change of the length of the arms for a fixed nanoribbon width w. In order to gain insight into the system behaviour, this was done by keeping the total ring length fixed. A change in the total of the upper arm, from $W/2$ to $W/2 + \Delta W$, led to a change in the lower, $W/2 \rightarrow W/2 - \Delta W$ (see upper right panel of Fig. 4.7). Interestingly, the position of the transmission peaks did not significantly change. This is a reasonable behaviour: as the total number of wavelengths which fit inside the ring is kept constant, the position with constructive interference in the ring is also fixed. On the other hand, Fano-like resonances show up between the maxima [20].

Another fabrication imperfection that could have a profound impact on transport is a change of the width of one of the arms. In this case, the lateral quantization in this wider (or narrower) section of the ring is different, resulting in a mode mismatch at the boundary, which can reduce the transmission. In order to estimate the corresponding effect, an extra layer of atoms was added to the horizontal section of the upper arm. The results are presented in Fig. 4.8, where the left panels show the transmissions $T_{\pm}$ for the asymmetric sample—filled lines—and the symmetric—solid black, for zero side-gate voltage, $U_G = 0$. As can be expected, the asymmetry shifts the transmission peaks. Besides, the overlaps between the transmission bands are reduced in the case of the asymmetric nanoring. These bands are more isolated, which results in the most important effect: the sign of the polarization can be switched much more abruptly. Therefore, the asymmetric design could be more advantageous for applications.

To conclude the discussion of the impact of fabrication imperfections on the device properties, it should be pointed out that, because the interference is very sensitive to many details, different samples could be expected to have quantitatively different current-voltage characteristics. However, as shown above, the qualitative behaviour

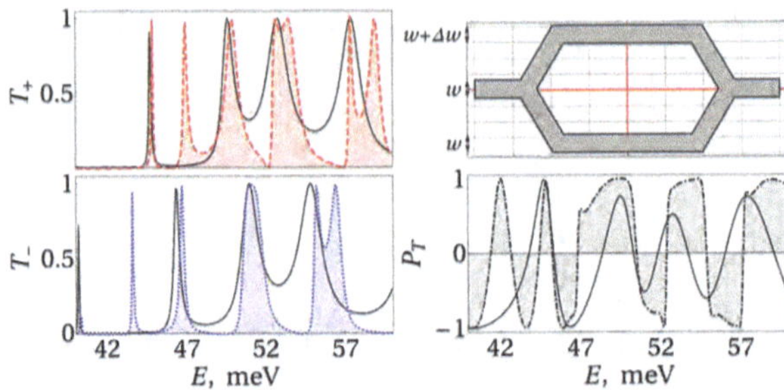

Fig. 4.8 Same as in Fig. 4.7, for an asymmetric nanoring engineered by changing the width of the *upper arm* to $N+1 = 64$ cells. This geometry leads to an enhanced, abruptly changing transmission polarization P_T

of the transmission coefficient and its polarization is robust under perturbations. It can then be expected that as long as the size of the ring, temperature, and applied voltages are such that the system remains within the one-mode regime, the device can operate as a tunable spin filter.

4.3.3 Origin of Fano Resonances in Asymmetric Rings

A general trend observed in the device is that the introduction of an asymmetry between the rings induces new resonances, characterized by points where the transmission exactly cancels. These resonances can be very sharp, as in Fig. 4.6, where couples of atoms were removed from the edges, or they can be comparable to the distance between transmission peaks, as in Fig. 4.8, where a geometric asymmetry was considered.

The origin of these resonances can be shown to be an intrinsic characteristic of systems having two scattering paths, each with a resonance at different energies. Under this condition, Fano resonances occur, thus showing up in a huge variety of physical systems, as reviewed in [21]. As a simple toy model to understand the effect of resonances in the systems, the ring sketched in Fig. 4.9 was used. It comprises two arms, each with two sites, connected to semi-infinite linear chains (the minimal model would only need to have one site per arm). The asymmetry has been included using a site dependent energy, $+\Delta\epsilon$ for the upper arm and $-\Delta\epsilon$ for the lower. Once again, the change in the energy of only one site is sufficient to induce an asymmetry, but the symmetric energy landscape in Fig. 4.9 was chosen to avoid shifts in the transmission when comparing different asymmetry strengths.

The resonant modes in the ring correspond to the eigenvalues of the subsystem comprised by the six sites in the ring. For $\Delta\epsilon = 0$, they can be straightforwardly calculated, leading to stationary waves with constant phase between the sites in the ring, $\theta_n = 2\pi n/6$, $n = 0, \ldots, 5$, shown in the left panel of Fig. 4.10. States labelled 1 and 4 are degenerate states with $E_1 = E_4 = -t$, and the same occurs for the states

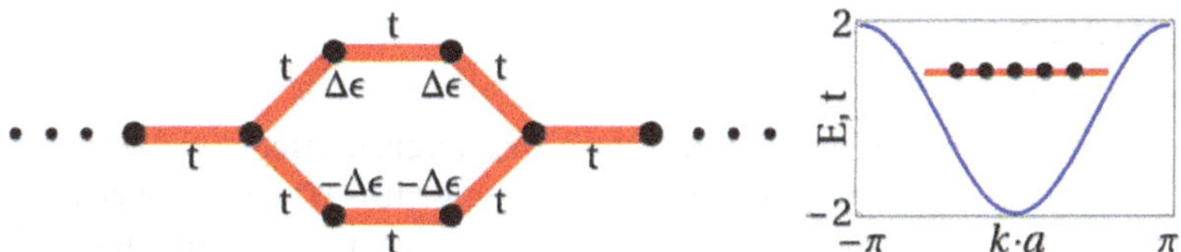

Fig. 4.9 The *left panel* shows the scheme of the system used to study the effects of asymmetries, whereas the dispersion relation of a linear chain with hopping parameter t is shown in the *right one*. The system comprises a ring with two arms, each with two sites, connected to semi-infinite linear chains. The labels $\pm\Delta\epsilon$ represent the energy shift with respect to the default $\epsilon_i = 0$

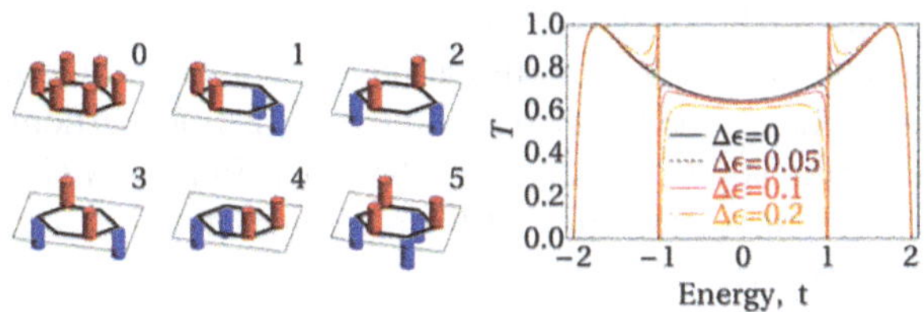

Fig. 4.10 *Left panel* eigenmodes of the symmetric ring ($\Delta\epsilon = 0$), with its index n given by the phase between neighbouring sites, $\theta_n = 2\pi n/6$. *Right panel* transmission through the ring for different values of the asymmetry parameter $\Delta\epsilon$, showing a strong shift at $E = t, 3t$, corresponding to the positions where the eigenmodes of the ring are degenerate in the symmetric case $\Delta\epsilon = 0$

2 and 3, with $E_2 = E_3 = t$. Finally, $E_0 = -2t$ and $E_5 = 2t$. It should be noted that the degeneracy in the eigenstates is a necessary condition to fulfil the invariance under rotations of the system.

The degeneracy in the eigenstates is broken when an asymmetry is introduced in the system, and this has a profound effect in the transmission for energies close to those points, as plotted in the right panel of Fig. 4.10. For any finite value of $\Delta\epsilon$, there is an energy value with transmission $T = 0$.

The typical width of the perturbation in the transmission scales with the strength of the asymmetry. This has a very simple physical meaning: if the asymmetry is small, the dephasing in the wave function after following both paths is very small. Therefore, if one needs to have a strong destructive interference (dephasing by π), the electrons need to stay a very long time inside the ring, so that it runs over the arms many times, and this only occurs when the energy is very close to the resonance. This is the case of the disorder in the edges of the rings, which mildly changes the effective wavelength of the wave function in the arms. On the other hand, strong asymmetries between the arms (see Figs. 4.7 and 4.8) relax the condition of being close to the resonance, as very few runs over the arms lead to a sufficient dephasing.

4.4 Summary

A novel spin filter which exploits quantum interference effects was proposed and studied. The device comprises a graphene nanoring with 60° bends fabricated above a ferromagnetic strip. It was shown that, due to the exchange splitting induced by the magnetic ions of the ferromagnetic layer, the transmission coefficient is different for spin up and spin down electrons, giving rise to the polarization of the conductance and the electric current. The ring geometry strongly enhances the current polarization, compared to a simple aGNR with a ferromagnetic layer on top of it, and it allows the current and its polarization to be controlled by a side-gate voltage.

A detailed study was made regarding the effect of edge disorder and other fabrication imperfections, such as the asymmetry of the ring. The predicted effects were shown to be robust under moderate perturbations. Under some circumstances, asymmetries in the ring enhance the behaviour of the polarization, particularly in the case where one of the arms is made wider than the other. Finally, a simple model was considered to explain the observation of Fano resonances in the transmissions $T_{\pm}$, as well as their relative width.

References

1. C.L. Kane, E.J. Mele, Quantum spin hall effect in graphene. Phys. Rev. Lett. **95**, 226801 (2005)
2. N. Tombros et al., Electronic spin transport and spin precession in single graphene layers at room temperature. Nature **448**, 571–574 (2007)
3. O.V. Yazyev, Hyperfine interactions in graphene and related carbon nanostructures. Nano Lett. **8**, 1011–1015 (2008)
4. E. Hill et al., Graphene spin valve devices. IEEE Trans. Magn. **42**, 2694–2696 (2006)
5. S. Cho et al., Gate-tunable graphene spin valve. Appl. Phys. Lett. **91**, 123105 (2007)
6. C. Józsa et al., Electronic spin drift in graphene field-effect transistors. Phys. Rev. Lett. **100**, 236603 (2008)
7. Y.S. Dedkov et al., Structural and electronic properties of Fe3O4/graphene/Ni(111) junctions. Phys. Status Solidi RRL **5**, 226–228 (2011)
8. Z.P. Niu, D.Y. Xing, Spin filter effect and large magnetoresistance in the zigzag graphene nanoribbons. Eur. Phys. J. B **143**, 139–143 (2010)
9. M. Ezawa, Spin filter, spin amplifier and other spintronic applications in graphene nanodisks. Eur. Phys. J. B **67**, 543–549 (2009)
10. F.S.M. Guimarães et al., Graphene-based spin-pumping transistor. Phys. Rev. B **81**, 233402 (2010)
11. F. Zhai, L. Yang, Strain-tunable spin transport in ferromagnetic graphene junctions. Appl. Phys. Lett. **98**, 062101 (2011)
12. A. Rozhkov et al., Electronic properties of mesoscopic graphene structures: Charge confinement and control of spin and charge transport. Phys. Rep. **503**, 77–114 (2011)
13. A.G. Swartz et al., Integration of the ferromagnetic insulator EuO onto graphene. ACS Nano **6**, 10063–10069 (2012)
14. H. Haugen et al., Spin transport in proximity-induced ferromagnetic graphene. Phys. Rev. B **77**, 115406 (2008)
15. J. Zou et al., Negative tunnel magnetoresistance and spin transport in ferromagnetic graphene junctions. J. Phys.: Condens. Matt. **21**, 126001 (2009)

16. Y. Gu et al., Equilibrium spin current in ferromagnetic graphene junction. J. Appl. Phys. **105**, 103711 (2009)
17. Y.-W. Son et al., Energy gaps in graphene nanoribbons. Phys. Rev. Lett. **97**, 216803 (2006)
18. A. H. Castro Neto et al. The electronic properties of graphene. Rev. Mod. Phys. **81**, 109–162 (2009)
19. A. Venugopal et al., Contact resistance in few and multilayer graphene devices. Appl. Phys. Lett. **96**, 013512 (2010)
20. U. Fano, Effects of configuration interaction on intensities and phase shifts. Phys. Rev. **124**, 1866–1878 (1961)
21. A.E. Miroshnichenko et al., Fano resonances in nanoscale structures. Rev. Mod. Phys. **82**, 2257–2298 (2010)

Chapter 5
Spin-Dependent Negative Differential Resistance in Graphene Superlattices

Since the pioneering work by Esaki in the late fifties [1], negative differential resistance (NDR) has been the principle of operation of many quantum devices. Usually, the underlying mechanism of the NDR is related to the resonant tunnelling of carriers through the device. As the chemical potential of a lead approaches to one of the resonant levels of the device, the current I increases. However, the resonant level position depends on the applied voltage V which can finally drive the system out of resonance. Then, the current can decrease dramatically with a further increase of the voltage. The resulting I–V characteristics are typically N-shaped with a region of NDR. Many devices display such non-monotonous I–V curves, for example, superconductor junctions [2], semiconductor superlattices [3], resonant tunnelling diodes [4], resonant interband tunnelling diodes [5], conductor/superconductor junctions [6], molecular films [7], carbon nanotubes [8], organic molecule/semiconductor junctions [9] and DNA molecules [10].

In this chapter, a spin-dependent superlattice, realized by ferromagnetic strips deposited on top of an Armchair Graphene Nanoribbon, is considered. The spin-dependent transmission and I–V curves are computed using both the full TB calculation and the approximate Dirac Hamiltonian. The two approaches are complementary: the former is more exact while the latter provides qualitative insight in the underlying physics. The model is similar to those studied by Niu et al. [11] and Ferreira et al. [12] but it goes beyond them in some important aspects. Gapless graphene with a periodic array of magnetic insulator strips was considered by Niu et al. They found that spin polarization of tunnelling conductance and magnetoresistance exhibit oscillatory behaviour as a function of the gate voltage. However, they did not take into account the finite width of the GNR and the quantization of the transverse momentum, which turns out to be an important effect. In the case of an aGNR, the quantized transverse momentum $k_\perp$ fixes the incidence angle $\widetilde{\theta} = \arcsin(k_\perp/k)$ as a function of the energy $\hbar k v_\mathrm{F}$, which is an essential feature of the system. Ferreira et al. [12] considered appropriate boundary conditions for the aGNRs and found NDR, but their superlattice is implemented by electrostatic gates and the NDR shows no spin dependence. Furthermore, the potential barriers required to obtain NDR in their system can not be attained using exchange interactions.

J. Munárriz Arrieta, *Modelling of Plasmonic and Graphene Nanodevices*,
Springer Theses, DOI: 10.1007/978-3-319-07088-9_5,

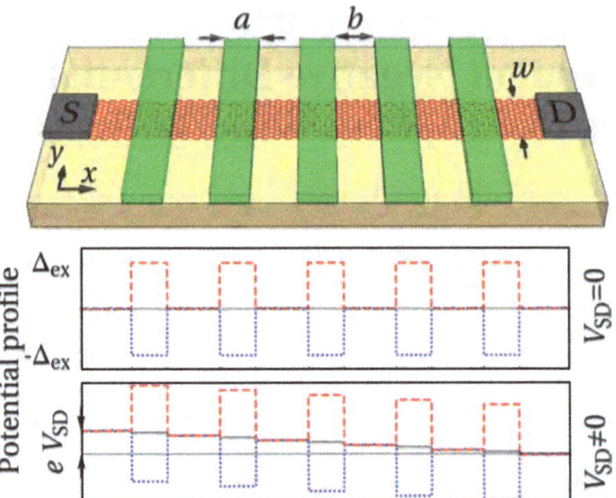

Fig. 5.1 *Upper panel* shows a schematic view of an aGNR with $N = 5$ strips of a ferromagnetic insulator placed on top of it (depicted using *green bars*). Source and drain leads are denoted as *S* and *D*, respectively. *Lower panels* show the spin-dependent potential profile in the unbiased ($V_{SD} = 0$, *middle panel*) and biased devices (*lower panel*). *Dashed red* and *dotted blue lines* show the potential profile for the spin-up and spin-down electrons, respectively. The *grey line* in the lower panel shows the electrostatic potential energy $\epsilon_{SD}(x)$ [see Eq. (5.1) in the text]

5.1 System and Modelling

The system under study is a GNR of width w connected to the source and drain leads. N strips of a ferromagnetic insulator of width a are arranged periodically on top of it, with spacing b between them, as shown schematically in the upper panel of Fig. 5.1. The choice of the width and the edge type of the GNR strongly affects its electronic properties. Here, GNRs with armchair edges, aGNRs, are used. Both experimental evidence [13] and *ab-initio* calculations [14] show that aGNRs present band gaps, which scale inversely with the width w of the GNR. The energy structure depends also on the remainder ($w/\tilde{a}_0 \mod 3$), where $\tilde{a}_0 = 0.246\,\text{nm}$ is the distance between next-nearest neighbours C atoms, namely the width of the hexagon in the honeycomb lattice. Moreover, contrary to the zGNRs, the dispersion relation is centred on $k = 0$. The conclusions of Chap. 3 suggest that this should be advantageous when the superlattice is considered, as the resonant levels are expected to be broader.

As described in Chap. 4, the strips of ferromagnetic insulator—EuO—lead to an effective Zeeman splitting of the spin sublevels, which results in spin-dependent potential profiles along the aGNR, as shown in the lower panels of Fig. 5.1, with two different energy level structures for electrons with opposite spins.

Contrary to the previous chapters, here it is necessary to go beyond the linear response theory to observe NDR, i.e., there must be a finite voltage drop along the ribbon. For simplicity, the source-drain voltage V_{SD} is assumed to drop at the edges of the ferromagnetic strips only, as shown by the grey line in the lower panel of Fig. 5.1. This is a reasonable assumption because other perturbations, such as functional groups, lead to measurable voltage drops [15]. However, a set of calculations was performed to check that the results do not depend crucially on the details of the potential profile: a modified model with V_{SD} falling homogeneously across the whole sample leads to very similar results.

5.1.1 Tight-Binding Hamiltonian and the QTBM

Analogously to the two previous chapters, the system was modelled following the recipe of Eq. (2.1). Only low-energy excitations were considered, and therefore the interactions between neighbouring atoms were restricted to the nearest ones, i.e., $l = 1$:

$$\mathcal{H} = -t \sum_{\langle i,j \rangle} |i\rangle \langle j| + \sum_{i} \epsilon_{\text{SD}}(x_i) \, |i\rangle \langle i| + \sigma \, \Delta_{\text{ex}} \sum_{i \in \mathcal{L}} |i\rangle \langle j| \; . \tag{5.1}$$

The on-site energy is the superposition of the following two terms: the bias-voltage induced electrostatic potential $\epsilon_{\text{SD}}(x_i)$, where x_i is the coordinate of the ith atom along the direction of the aGNR (grey line in the lower panel of Fig. 5.1), and the exchange-interaction shift $\sigma \Delta_{\text{ex}}$ induced by the ferromagnetic strips, introduced following the guidelines of Sect. 4.1. A spin-up (down) electron propagating along the sample is subjected to a potential comprising a set of rectangular barriers (wells), as plotted in the middle panel of Fig. 5.1. Hereafter, $\Delta_{\text{ex}} = 5\,\text{meV}$ is used, which lies in the range of values known from the literature, 3–10 meV [16–18]. It should be noted that different choices for this parameter lead to similar results if the ribbon width and/or strips geometry is changed accordingly.

Using the Landauer-Büttiker scattering formalism presented in Sect. 2.3, the electric current across the sample was calculated, following the method detailed in Appendix A. This allowed the spin-dependent transmission coefficients $T_{\pm}$ for spin-up (+) and spin-down (−) electrons, and the spin-polarized currents, $I_{\pm}$, to be calculated.

5.1.2 Dirac Theory for aGNRs

For not-too-narrow aGNRs ($w \gtrsim 3\,\text{nm}$), it is possible to compute the transmission coefficient very efficiently using the Dirac theory. For energies close to the Dirac points, electrons in graphene can be described effectively by a two-dimensional two-valley Dirac equation [19, 20]. The boundary conditions of aGNRs require the wave function to vanish on the (hypothetical) sites just outside the aGNR, i.e., at $y = 0$ and $y = w + \tilde{a}_0$, with $\tilde{a}_0 \equiv \sqrt{3}a_0$. The lower edge of the aGNR is located at $y = \tilde{a}_0/2$ (see Fig. 5.2). In the case of armchair aGNRs, this affects both sublattices

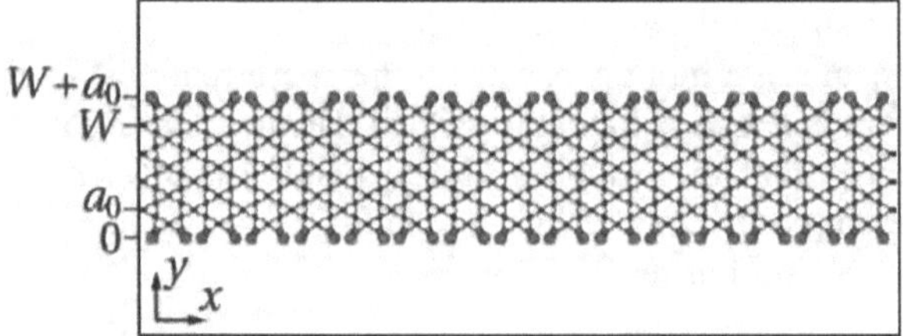

Fig. 5.2 Scheme of a ribbon, with *white disks* and *black lines* representing real sites and bounds, respectively. *Grey disks* and *lines* represent the hypothetical sites just outside the aGNR, added to set the wave function to 0 in them and fulfil the boundary conditions. The heights of consecutive rows are $\tilde{a}_0/2$ apart, and range from $y = 0$ to $y = w + \tilde{a}_0$, with the extremes corresponding to the hypothetical sites

and the boundary conditions can be fulfilled by a superposition of two states from different valleys with the same energy $E = \hbar v_F (k_\perp^2 + k_\parallel^2)^{1/2}$ and equal longitudinal momentum $\hbar k_\parallel$, but with opposite transverse momentum $\pm\hbar k_\perp$, measured from the Dirac points [21, 22].

5.1.2.1 Quantization of Transverse Momentum

Since the valley momentums $\boldsymbol{K}$ and $\boldsymbol{K}'$ can be chosen parallel to $\boldsymbol{k}_\perp$,

$$\boldsymbol{K} = \left(0, -\frac{4\pi}{3\tilde{a}_0}\right) \qquad \boldsymbol{K}' = \left(0, \frac{4\pi}{3\tilde{a}_0}\right), \tag{5.2}$$

the transverse wave function can be written $\phi_\perp(y) = \sin\big[(K + k_\perp)y\big]$ where $K = 4\pi/(3\tilde{a}_0)$. This function is evaluated on the honeycomb lattice with $2y/\tilde{a}_0 \in \mathbb{N}$ and oscillates rapidly. The transverse momentum $k_\perp$ is quantized by the conditions $\phi_\perp(w + \tilde{a}_0) = \phi_\perp(0) = 0$. The allowed values for $k_\perp$ are given by $(K + k_{\perp n})(w + \tilde{a}_0) = n\pi$, $n \in \mathbb{Z}$, and the spectrum is $E_n(k_\parallel) = \pm\hbar v_F\sqrt{k_{\perp n}^2 + k_\parallel^2}$.

Taking into account that w is an integer multiple of $\tilde{a}_0/2$, one finds that the spectrum is gapless if [22],

$$w = (3n' + 1)\tilde{a}_0/2\,, \quad n' \in \mathbb{N}\,. \tag{5.3}$$

For asymmetric aGNR, as in [21], n' is even. For symmetric aGNR, w is an integer multiple of $\tilde{a}_0$ and n' is odd, such that $w = (3n - 1)\tilde{a}_0$, $n \in \mathbb{N}$ implies a gapless spectrum. This result agrees with the results obtained for Tight Binding calculations and presented in Sect. 2.2.2, where it is also explained that in real samples there are small gaps even in the cases which fulfil Eq. (5.3), which are due to edge effects not included in the simple Dirac ansatz nor the homogeneous tight-binding formulation [13, 14]. In the following, the only ribbons that are taken into account are symmetric aGNRs of width $w = n''\tilde{a}_0$, where the integer n'' is different from $3\mathbb{N} - 1$,

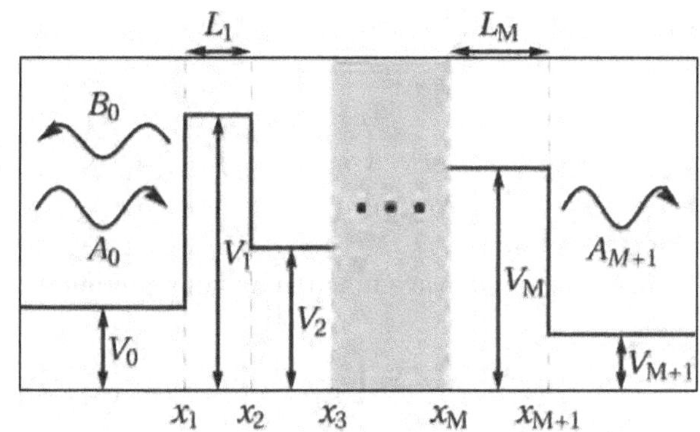

Fig. 5.3 Transmission across a series of M potential steps. The incident plane wave with amplitude A_0 splits into a reflected and a transmitted component with amplitudes B_0 and A_{M+1}, respectively

e.g., $w = 40\tilde{a}_0$. These have a gap anyway and are quite robust against edge effects. Then, the allowed values of the transverse momentum are

$$|k_{\perp n}| = \frac{\pi n}{3(w + \tilde{a}_0)}, \quad n = 1, 2, 4, 5, 7, 8, \ldots, \tag{5.4}$$

and the half-gap is $E_0 = E_1(0) = \hbar v_F \pi/[3(w+\tilde{a}_0)]$. For energies E from the range $[E_1(0), E_2(0)]$, only the lowest transverse mode with $k_{\perp 1}$ can propagate.

5.1.2.2 Transfer-Matrix Description of Transmission

For geometries like the one shown in Fig. 5.1, where the (spin-dependent) potential depends only on the longitudinal coordinate x, the transverse momentum $k_\perp$ together with the wave function $\phi_\perp(y)$ is conserved, and it suffices to solve for the longitudinal wave function $\phi_\parallel(x)$. Then, it is necessary to calculate the transmission across a piecewise constant potential profile, as sketched in Fig. 5.3. The solution of the Dirac equation for each spin $\sigma = \pm 1$ and in each interval of constant potential value V is the superposition of two counter-propagating sublattice pseudo-spinors

$$\psi_\parallel(x) = A \begin{pmatrix} e^{-i\theta/2} \\ -e^{i\theta/2} \end{pmatrix} e^{ik_\parallel x} + B \begin{pmatrix} e^{+i\theta/2} \\ -e^{-i\theta/2} \end{pmatrix} e^{-ik_\parallel x}, \tag{5.5}$$

with $\tan\theta = k_\parallel/k_\perp$ and $k_\parallel = [(E-V)^2/(\hbar v_F)^2 - k_\perp^2]^{1/2}$. The solution may be evanescent because Eq. (5.5) holds also for $|E-V| < \hbar v_F |k_\perp|$, when $k_\parallel$ and θ become imaginary. The general form of the wave function in each slab j with potential V_j and momentum $k_\parallel = k_j$ then is

$$\begin{pmatrix} e^{-i\theta_j/2} & e^{i\theta_j/2} \\ -e^{i\theta_j/2} & -e^{-i\theta_j/2} \end{pmatrix} \begin{pmatrix} A_j(x) \\ B_j(x) \end{pmatrix} \equiv S_j \begin{pmatrix} A_j(x) \\ B_j(x) \end{pmatrix}, \tag{5.6}$$

where $A_j(x) = Ae^{ik_jx}$ and $B_j(x) = Be^{-ik_jx}$, such that

$$\begin{pmatrix} A_j(x_{j+1}) \\ B_j(x_{j+1}) \end{pmatrix} = G_j \begin{pmatrix} A_j(x_j) \\ B_j(x_j) \end{pmatrix}, \tag{5.7}$$

with $G_j = \text{diag}\big(\exp(ik_jL_j), \exp(-ik_jL_j)\big)$ and $L_j = x_{j+1} - x_j$. At each junction, k_j changes, but the wave function remains continuous,

$$S_j \begin{pmatrix} A_j(x_j) \\ B_j(x_j) \end{pmatrix} = S_{j-1} \begin{pmatrix} A_{j-1}(x_j) \\ B_{j-1}(x_j) \end{pmatrix}. \tag{5.8}$$

Then, the transfer matrix for the whole system can be written with the help of Eqs. (5.7) and (5.8)

$$\begin{pmatrix} A_{M+1} \\ B_{M+1} \end{pmatrix} = S_{M+1}^{-1} \widetilde{G}_M \ldots \widetilde{G}_2 \widetilde{G}_1 S_0 \begin{pmatrix} A_0 \\ B_0 \end{pmatrix}, \tag{5.9}$$

with $\widetilde{G}_j = S_j G_j S_j^{-1}$.

For the transmission problem depicted in Fig. 5.3, the boundary condition is no incident current from the right, $B_{M+1} = 0$. The reflection coefficient at the left is the ratio of reflected to incident current, $R = |B_0|^2/|A_0|^2$. For the transmission coefficient one has to take into account that the longitudinal momentums k_{M+1} and k_0 are different if $V_0 \neq V_{M+1}$, such that the ratio of transmitted to incident current is $T = (|A_{M+1}|^2 k_{M+1})/(|\tilde{a}_0|^2 k_0)$.

5.1.2.3 Band Structure of the Unbiased Lattice

For an unbiased lattice with identical barriers of width l_1 and spacing l_0, there are only two different transfer matrices involved, $\widetilde{G}_0$ and $\widetilde{G}_1$. In the limit $N \to \infty$, the superlattice eigenfunctions have the Bloch phases $\exp(\pm i K_x l)$ that are the eigenvalues of the transfer matrix $\widetilde{G} = \widetilde{G}_0 \widetilde{G}_1$ over one lattice period $l = l_0 + l_1$,

$$\widetilde{G} = \frac{1}{\sin(\theta_1)\sin(\theta_2)} \begin{pmatrix} \mathcal{S}(-\theta_1, -\theta_2; 0, 0) & \mathcal{S}(0, -\theta_2; \theta_1, 0) \\ \mathcal{S}(0, \theta_2; -\theta_1, 0) & \mathcal{S}(\theta_1, \theta_2; 0, 0) \end{pmatrix}, \tag{5.10}$$

with

$$\mathcal{S}(\alpha_1, \alpha_2; \alpha_3, \alpha_4) \equiv \sin(k_1 l_1 + \alpha_1)\sin(k_2 l_2 + \alpha_2) - \sin(k_1 l_1 + \alpha_3)\sin(k_2 l_2 + \alpha_4).$$

The dispersion relation $E(K_x, K_y = k_\perp)$ is calculated by solving the eigenvalue problem $\widetilde{G}\psi_{\parallel,K_x} = e^{iK_x l}\psi_{\parallel,K_x}$, which leads to the implicit relation $\cos(K_x l) = \text{Tr}(\widetilde{G})/2$, or again [23, 24]

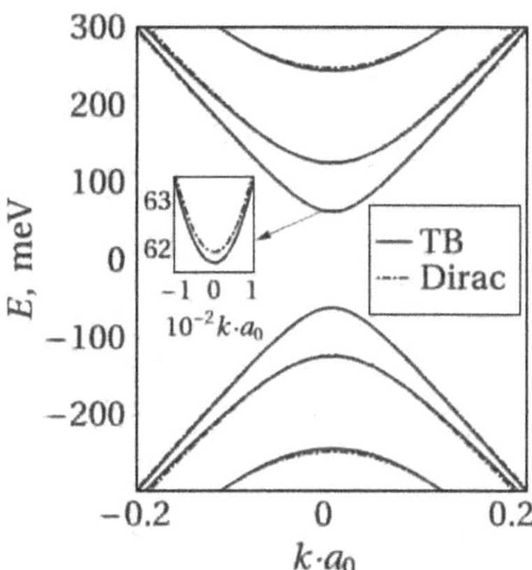

Fig. 5.4 Dispersion relation of a nanoribbon of width $w = 40\tilde{a}_0$, calculated using the TB Hamiltonian (*solid lines*) and the Dirac approximation (*dot-dashed lines*). The *inset* shows the dispersion relation near the band edge, showing the small shift due to the assumptions of the Dirac approximation

$$\cos(K_x l) = \cos k_0 l_0 \cos k_1 l_1 + \frac{\cos\theta_0 \cos\theta_1 - 1}{\sin\theta_0 \sin\theta_1} \sin k_0 l_0 \sin k_1 l_1. \tag{5.11}$$

If $|\,\mathrm{Tr}(\widetilde{G})/2| > 1$, there is no propagating solution with $K_x \in \mathbb{R}$, and E falls into the band gap of the superlattice. In Fig. 5.5 below, the bands of the infinite superlattice are shown together with the transmission across a finite sample of five barriers. Physically, this bands stem from the resonant levels placed in the regions with lowest potentials, which, for the potential steps $V_0 < V_1$, are given by the expression $(\widetilde{G}_0)_{2,2} = 0$.

5.2 Results and Discussion

The sample considered consists of an aGNR of width $w = 40\tilde{a}_0 \simeq 9.84$ nm and $N = 5$ ferromagnetic strips of width $a = 23.9$ nm and inter-strip spacing $b = 55.8$ nm (Fig. 5.1 shows a schematic view of the device). For this width, the dispersion relation of the ribbon shows a half-gap $E_0 = 61.7$ meV.

5.2.1 Transmission for Zero Bias

First, the transmission of electrons through the sample was calculated, using the two methods outlined in Sect. 5.1. In the upper panel of Fig. 5.5, the transmissions $T_\pm$ are shown for unbiased aGNRs ($V_{SD} = 0$). Comparing the results of the TB to those of the Dirac model, a very good agreement is found, which validates the Dirac theory applied to aGNRs. It should be noted that the Dirac model slightly overestimates the band gap ($E_0 = 61.9$ meV; see Fig. 5.4). As detailed in [21], this is due to the

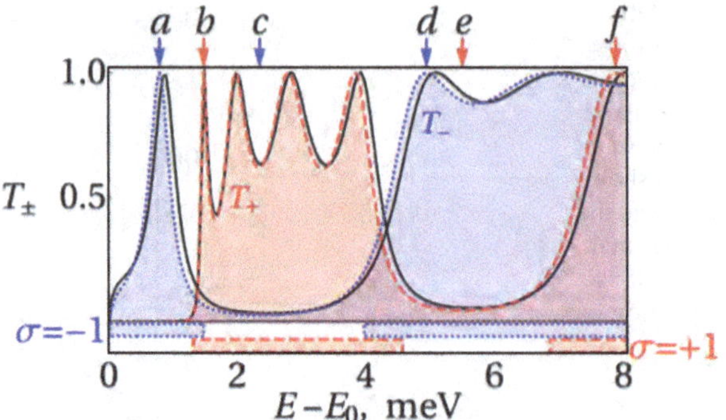

Fig. 5.5 Transmission coefficients as a function of energy for the unbiased device with $N = 5$ strips. There is a very good agreement between the TB calculation (*solid black lines*) and Dirac transfer-matrix theory (*filled lines*). The letters *a–f* in the upper part of the figure label the energies with particularly high or low transmission; the corresponding wave functions are plotted in Fig. 5.6. The horizontal bars in the lower part of the figure indicate the bands of the infinite superlattice, calculated using Eq. (5.11)

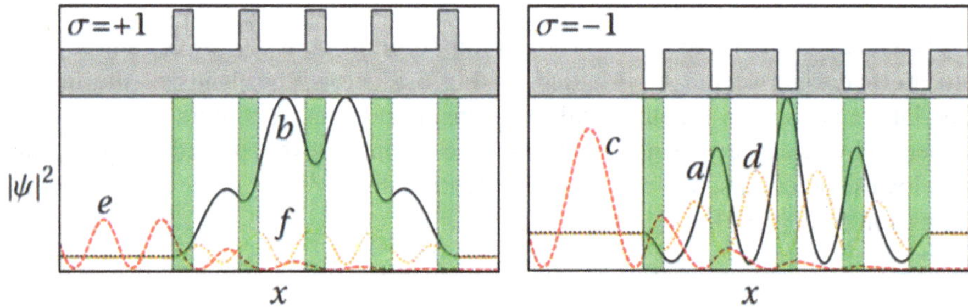

Fig. 5.6 Probability density for the states with energies marked by letters *a–f* and *red* (*blue*) *arrows* for $\sigma = +1$ ($\sigma = -1$) in Fig. 5.5. In the resonant cases with full transmission (*solid black* and *dotted orange*), the in- and out-going waves couple to a quasi-bound state, which manifests itself in a typical concentration of the density in the scattering region (i.e., within the superlattice). In the cases of strong reflection (*dashed red*), the probability density decays across the sample

$\boldsymbol{k} \cdot \boldsymbol{p}$ approximation in the derivation of the Dirac equation, which breaks down for very narrow aGNRs. In Fig. 5.5, the energy is measured from the respective band edges for each data set, which eliminates the shift of E_0 from the plots. Since the transfer-matrix calculation in Eq. (5.9) requires much less computational resources than the full TB calculation, only the Dirac theory is used in the following.

The transmission strongly depends both on the energy E of the injected carrier and its spin. The obtained transmission pattern is typical for a superlattice. Already for $N = 5$, the regions of high transmission coincide quite well with the bands of the infinite superlattice obtained in Eq. (5.11) and given by the horizontal bars at the bottom of the figure. As illustrated in Fig. 5.6, the states with full transmission appear due to the resonant coupling of the propagating waves to quasi-bound states of the superlattice.

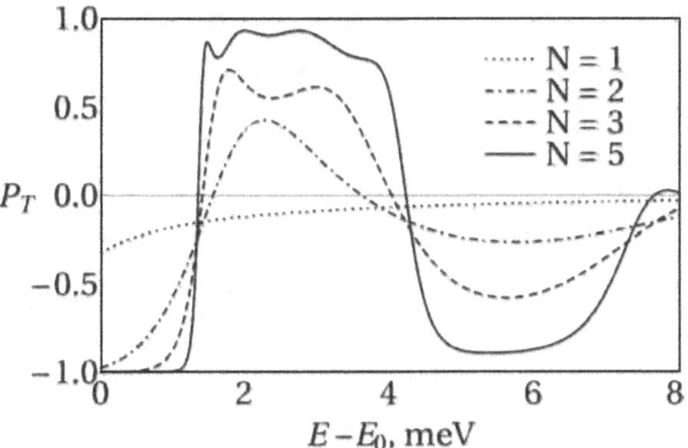

Fig. 5.7 Transmission polarization P_T, given by Eq. (2.12), as a function of energy for $V_{SD} = 0$ and different numbers N of ferromagnetic strips. For $N \gtrsim 3$, the curves manifest abrupt polarization switching within narrow energy intervals

Due to the level repulsion, these regions with high transmission have in general $N - 1$ peaks in the case of a superlattice of barriers, and N in the case of wells. However, the first transmission band for $\sigma = -1$ shows only one maximum. This is due to the fact that the corresponding band of the infinite superlattice is placed at $[E_0 - 1.71\,\text{meV},\ E_0 + 1.47\,\text{meV}]$, and therefore it falls only partially above the subband edge.

The energies of the transmission peaks and transmission bands depend on the spin σ, which manifests itself clearly in the transmission polarization P_T, defined in Eq. (2.12). The latter quantity is shown in Fig. 5.7. When the system parameters are chosen such that the overlap between transmission bands corresponding to different spins is small, P_T presents abrupt switching from -1 to 1, and vice versa, within narrow energy intervals. The energy of band edges is only slightly affected by the number of barriers N, as long as a and b are kept fixed. But as N is increased, the transmission of modes with energies outside the transmission bands tends rapidly to 0, thus leading to an enhanced transmission polarization. This is shown in Fig. 5.7, for $N = 1, 2, 3, 5$. Interestingly, the first transmission band for $\sigma = -1$ (vertical arrow labelled a in Fig. 5.5) only appears if $N \geq 3$.

5.2.2 Currents and Spin Polarization for Finite Bias

Next, the current response of the device to a bias V_{SD} between source and drain is described. The chemical potentials in the leads, $\mu_S = eV_{SD} + \mu$ and $\mu_D = \mu$ have the same offset μ from their respective band-gap centres, resulting in a piecewise constant potential profile as shown in the lower panel of Fig. 5.1, and discussed in Sect. 5.1. The transmission coefficients $T_\pm$ as functions of E and V_{SD} are computed with the transfer-matrix description. The bias then results in a distortion of the transmission bands shown in Fig. 5.5. Basically, the transmission bands move towards lower energies and finally go away (see Fig. 5.8a–b).

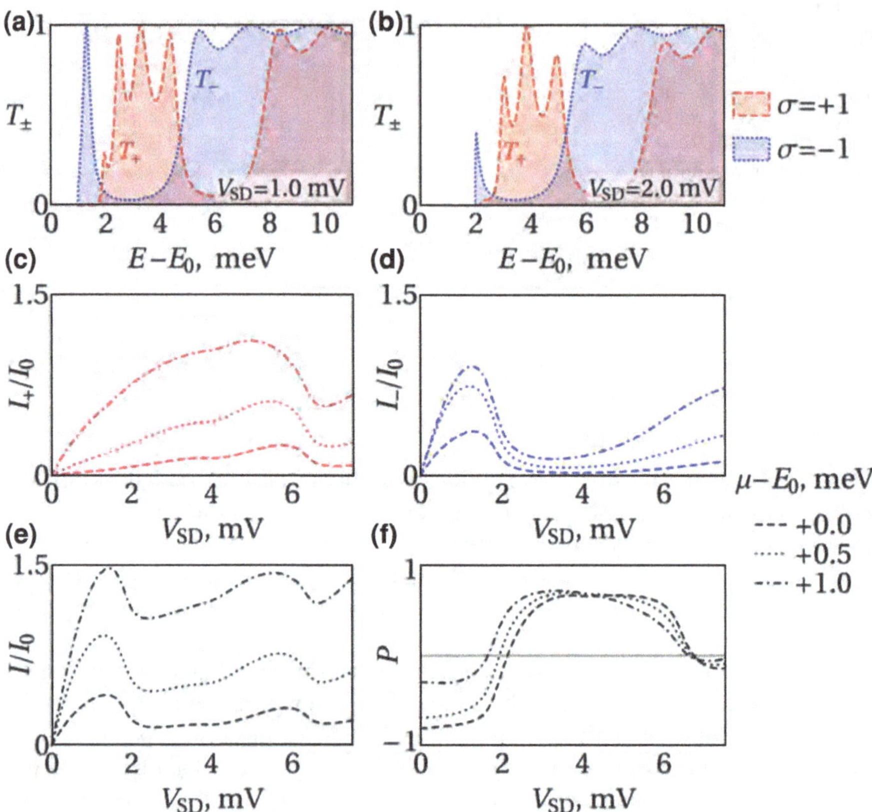

Fig. 5.8 Transmission and currents for finite bias. *Panels a* and *b* show that the transmissions for both spins with finite bias V_{SD} are shifted and distorted compared to Fig. 5.5. *Panels c* and *d* show the currents $I_\pm$ at $T = 4\,\text{K}$ in units of $I_0 = 10^4 e/(ht)$ as function of V_{SD} for different values of the chemical potential μ. *Panels e* and *f* represent the total current I and current polarization P for the same parameters

Within the Landauer-Büttiker scattering formalism, the spin-polarized currents $I_\pm$ are obtained using Eq. (2.13). The currents $I_\pm$ as function of V_{SD} are plotted in the panels c and d of Fig. 5.8. The spin-dependent transmission bands and the distortion due to the bias lead to regions of NDR at different values of the bias voltage for the different spins. For spin down ($\sigma = -1$), the NDR occurs at a low bias voltage and its slope is particularly steep. This is due to the fact that the single transmission peak labelled *a* in Fig. 5.5 is still very sharp when it suddenly disappears. The lowest transmission band of spin up ($\sigma = +1$) disappears only at a rather high bias voltage, when its profile has already become washed out. This is the reason why the NDR of $\sigma = -1$ is much sharper than the one of $\sigma = +1$.

It is also interesting to calculate the properties of the current due to initially unpolarized charge carriers. Therefore, the total current through the device I, defined in Eq. (2.14), as well as its output polarization P, taken from Eq. (2.15), were calculated. These quantities are plotted in the panels e and f of Fig. 5.8. The polarization shows an initial range with negative values, followed by a second region dominated by spin-up electrons, until the system reaches an unpolarized regime. Regarding the total current, regions with NDR are also found. This is due to the slope in the NDR region of I_- (I_+) being much steeper than the positive slope of the I_+ curve (I_- curve).

5.3 Effect of Geometric Disorder

The properties of the considered system heavily rely on the geometry of the superlattice. However, deviations from the values set for a and b are to be expected in the fabrication of the device. Therefore, it is important to find out if the loss of the regularity in the sample destroys the predicted effect, or these are robust under moderate perturbations of the geometry.

Contrary to previous chapters, the Dirac formalism developed here allows the effect of disorder to be calculated very efficiently using transfer matrices, provided that the strips remain perpendicular to the aGNR. Therefore, the properties of a large number of samples can be studied within relatively low computational time. As discussed previously, the separation of the transmission bands for different spins is the main ingredient to obtain a spin-polarized current with good characteristics. Therefore, in order to study the *ensemble*, it is necessary to define a fitness function which allows these separations to be quantified. The following function is proposed:

$$W(T_\pm) = \frac{1}{E_{\max}} \int_{E_0}^{E_0+E_{\max}} dE \, |T_+ - T_-|^2 \, . \tag{5.12}$$

The value $E_{\max} = 6\,\mathrm{meV}$ was chosen, because this is the position of the minimum for T_+ in the ordered sample, after its corresponding transmission band. The functional form of $W(T_\pm)$ was chosen so that strong overlaps between $T_\pm$ or low values for the transmission lead to $W(T_\pm) \simeq 0$.

A very general type of geometric disorder was considered, with the strip width at the ith cell, a_i, and inter-spacing b_i, independently varying by up to 10 % of the total cell width $l = a + b$. For the parameters chosen in the ordered configuration, $a_i \in [0.25l, 0.35l]$, and $b_i \in [0.65l, 0.75l]$. The middle panel of Fig. 5.9 shows the probability distribution of $W(T_\pm)$, with the value of the ordered configuration marked with a solid black line. The average value of $W(T_\pm)$ shifts towards lower values. However, an important subset of the *ensemble* shows an enhancement of the transmission properties. The main mechanism contributing to this enhancement is the broadening of the first transmission band in T_-, as exemplified in the lower right panel of Fig. 5.9.

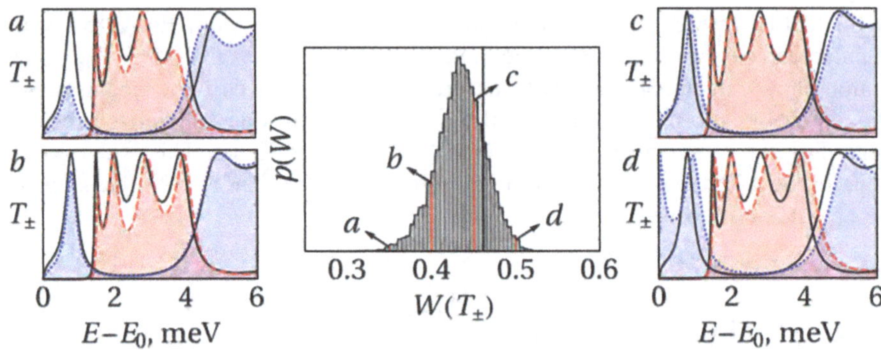

Fig. 5.9 Effect of geometric disorder on the transmission. The *middle panel* shows the distribution probability of $W(T_\pm)$, for the strip widths a_i and inter-spacings b_i varying up to $l/20$ around the values of the ordered configuration. The value for the ordered configuration is marked using a *solid black line*. The transmissions $T_\pm$ for four particular configurations, labelled *a–d*, are shown in the *left* and *right columns*, with *black lines* representing the ordered case

5.4 Summary

In summary, a novel graphene-based device comprising an aGNR and a regular array of ferromagnetic strips grown on top of it has been proposed and studied. The strips introduce a short-range exchange splitting of electronic states in the aGNR and create a spin-dependent superlattice. It has been shown that the electric current through the device can be highly polarized. Moreover, the two polarized components of the current manifest non-monotonic dependencies on the source-drain voltage. In particular, both spin-dependent current-voltage characteristics present regions with negative differential resistance for the source-drain voltage in the range of a few millivolts. The device operates therefore as a low-voltage regular Esaki diode for the spin-polarized currents.

Moreover, there are several additional advantages of the proposed device design. In particular, the usage of a superlattice induced by ferromagnets rather than by usual electrostatic gates is very attractive from the point of view of the circuit integration and device density. Unlike the long-range electrostatic gate potentials which can interfere with each other, setting a practical lower limit for the inter-device distance, the exchange interaction is very short-ranged. Its characteristic length scale is on the order of one monolayer, which allows for very close packing of circuits and, consequently, considerably higher device densities.

Finally, it should be noted that in a true spintronic device the degree of freedom that carries information is the polarization of the current rather than its magnitude. It was shown that the current polarization is also a non-monotonic function of the source-drain voltage in our proposed device, which makes it an Esaki spin diode. The latter opens a possibility to design a whole new class of true spintronic circuits such as spin oscillators, amplifiers and triggers.

References

1. L. Esaki, New phenomenon in narrow germanium p-n junctions. Phys. Rev. **109**, 603–604 (1958)
2. I. Giaever, Electron tunneling between two superconductors. Phys. Rev. Lett. **5**, 464–466 (1960)
3. R. Tsu, L. Esaki, Tunneling in a finite superlattice. Appl. Phys. Lett. **22**, 562–564 (1973)
4. L.L. Chang et al., Resonant tunneling in semiconductor double barriers. Appl. Phys. Lett. **24**, 593 (1974)
5. J.R. Soderstrom et al., New negative differential resistance device based on resonant interband tunneling. Appl. Phys. Lett. **55**, 1094–1096 (1989)
6. R. Kümmel et al., Andreev scattering of quasiparticle wave packets and current-voltage characteristics of superconducting metallic weak links. Phys. Rev. B **42**, 3992–4009 (1990)
7. Y. Xue et al., Negative differential resistance in the scanning-tunneling spectroscopy of organic molecules. Phys. Rev. B **59**, 7852 (1999)
8. F. Léonard, J. Tersoff, Negative differential resistance in nanotube devices. Phys. Rev. Lett. **85**, 4767–4770 (2000)
9. S.Y. Quek et al., Negative differential resistance in transport through organic molecules on silicon. Phys. Rev. Lett. **98**, 066807 (2007)
10. A.V. Malyshev, DNA double helices for single molecule electronics. Phys. Rev. Lett. **98**, 096801 (2007)
11. Z.P. Niu et al., Spin transport in magnetic graphene superlattices. Eur. Phys. J. B **66**, 245–250 (2008)
12. G.J. Ferreira et al., Low-bias negative differential resistance in graphene nanoribbon superlattices. Phys. Rev. B **84**, 125453 (2011)
13. M.Y. Han et al., Energy band-gap engineering of graphene nanoribbons. Phys. Rev. Lett. **98**, 206805 (2007)
14. Y.-W. Son et al., Energy gaps in graphene nanoribbons. Phys. Rev. Lett. **97**, 216803 (2006)
15. L. Yan et al., Local voltage drop in a single functionalized graphene sheet characterized by kelvin probe force microscopy. Nano Lett. **11**, 3543–3549 (2011)
16. H. Haugen et al., Spin transport in proximity-induced ferromagnetic graphene. Phys. Rev. B **77**, 115406 (2008)
17. J. Zou et al., Negative tunnel magnetoresistance and spin transport in ferromagnetic graphene junctions. J. Phys.: Condens. Matter **21**, 126001 (2009)
18. Y. Gu et al., Equilibrium spin current in ferromagnetic graphene junction. J. Appl. Phys. **105**, 103711 (2009)
19. P.R. Wallace, The band theory of graphite. Phys. Rev. **71**, 622–634 (1947)
20. A.H. Castro Neto, F. Guinea, Impurity-induced spin-orbit coupling in graphene. Phys. Rev. Lett. **103**, 026804 (2009)
21. L. Brey, H.A. Fertig, Electronic states of graphene nanoribbons studied with the Dirac equation. Phys. Rev. B **73**, 235411 (2006)
22. K. Wakabayashi et al., Electronic transport properties of graphene nanoribbons. New J. Phys. **11**, 095016 (2009)
23. M. Barbier et al., Extra Dirac points in the energy spectrum for superlattices on single-layer graphene. Phys. Rev. B **81**, 075438 (2010)
24. Q. Zhao et al., Localization behavior of Dirac particles in disordered graphene superlattices. Phys. Rev. B **85**, 104201 (2012)

Part II
Electro-Optical Nanodevices

Chapter 6
Optical Nanoantennas with Tunable Radiation Patterns

Radio and microwave antennas operate as receivers or emitters of electromagnetic radiation within the corresponding wavelength ranges. The size of these devices is typically on the order of the wavelength, which enables one to convert the freely propagating electromagnetic energy into a localized excitation of the antenna and vice versa. Fabricating an optical antenna requires technology of producing objects with sub-wavelength size, down to several tens of nanometres, which is now possible with modern methods of nanotechnology [1–6].

Recently, plasmonic antennas, operating in the visible range of the spectrum, have received a great deal of attention [7–9]. Among others, resonator [10], bow-tie [11, 12], Yagi-Uda [13], graded [14, 15], cross-resonant [16], core-shell [17] and nanorod [18] configurations have been investigated. Nonlinear plasmonics [19], nanoantenna-enhanced gas sensing [20] as well as nanoscale spectroscopy [21] with optical antennas have also been discussed.

Usually, plasmonic antennas comprise arrays of metallic nanoparticles or nanowires, which convert propagating optical signals into surface plasmon modes or vice versa. One of the challenging tasks here is the antenna excitation. Several schemes have been proposed, ranging from excitation by an adjacent point emitter (such as a single molecules or quantum dots) [18], which requires a very high precision of positioning of the point source, to the excitation by a beam of an electron microscope [13]. Other important and challenging aspects are the design and control of the radiation patterns of such antennas [13, 18]. In this chapter both issues are addressed theoretically. An antenna, consisting of a regular linear array of metal nanospheres located close to the interface of two materials with high dielectric contrast, is considered. It is shown that the radiation pattern of such a system can be controlled and tuned in a variety of ways, in particular, by changing the angles of incidence and polarization of the excitation beam. The system is illuminated using evanescent waves, which is advantageous from the view point of separation of the excitation from the antenna signal. The radiation pattern of the considered antenna is strongly directional and highly sensitive on the excitation parameters, which can be

J. Munárriz Arrieta, *Modelling of Plasmonic and Graphene Nanodevices*,
Springer Theses, DOI: 10.1007/978-3-319-07088-9_6,

explained by the interference between the field created by the nanoparticle electric dipoles and their images induced by the interface.

6.1 Model and Formalism

The system considered comprises an array of MNPs in close proximity to the interface of two dielectrics with permittivities ϵ_1 and $\epsilon_2 > \epsilon_1$ (see Fig. 6.1). The array is a linear chain of N equally spaced identical metal spheres with radii a and centre-to-centre distances d, which are embedded in a dielectric with permittivity ε_1 at a centre-to-interface distance h (the chain is parallel to the interface). The ratios $a/d \leq 1/3$ and $a/h \leq 1/3$ are chosen in such a way that the point dipole approximation for the interactions between different particles can be used. The former condition on the sphere radius and the centre-to-centre distance was discussed in [22, 23]. The latter relationship is analogous; to verify its validity, results obtained within the point dipole approximation were compared with those calculated using the boundary-elements method [24], which were in good agreement. As methods based on the point dipole approximation use simpler and numerically less demanding, they will be used henceforth.

The nanospheres are characterized by their polarizabilities $\alpha(\omega)$, where ω is the frequency of the incident field. The polarizability of small particles (compared with the wavelength) in a homogeneous environment can be described retaining the first two correction terms [25, 26]:

$$\frac{1}{\alpha(\omega)} = \frac{1}{\alpha^{(0)}(\omega)} - \frac{k_1^2}{a} - i\,\frac{2}{3}\,k_1^3\,, \tag{6.1}$$

where $\alpha^{(0)}(\omega)$ is the bare quasi-static polarizability of the sphere, $k_1 = (\omega/c)\sqrt{\epsilon_1}$ is the wave number of light in the medium embedding the spheres, and c is the speed of light in vacuum. The term quadratic in k_1 describes the depolarization shift of the plasmon resonance, while the cubic one accounts for the radiative damping [26]. Dielectric or metallic surfaces in the proximity of a sphere can also modify its polarizability (see below). Within the quasi-static approximation, the bare polarizability $\alpha^{(0)}$ is expressed in terms of the frequency dependent dielectric constant of the bulk nanoparticle material, $\varepsilon_M(\omega)$, and that of the embedding dielectric [27, 28]:

$$\alpha^{(0)}(\omega) = a^3\,\frac{\varepsilon_M(\omega) - \varepsilon_1}{\varepsilon_M(\omega) + 2\,\varepsilon_1}\,. \tag{6.2}$$

In the point dipole approximation, the induced dipole moment of the n-th sphere $\mathbf{p}_n$ can be obtained by solving the following set of equations:

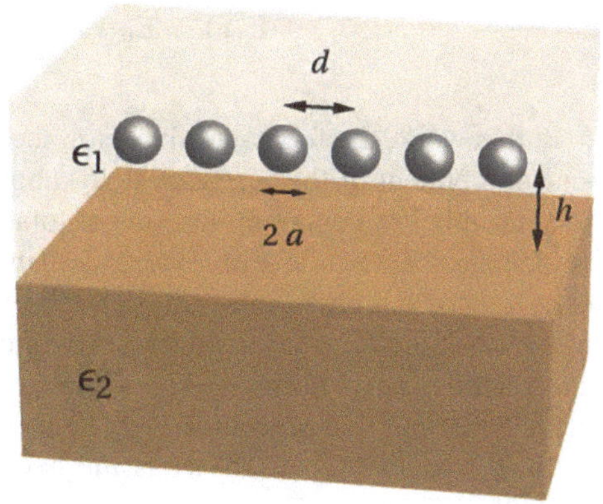

Fig. 6.1 System schematics: a regular linear array of identical spherical MNPs located parallel to the interface between two dielectrics with permittivities ϵ_1 and ϵ_2 ($\epsilon_1 < \epsilon_2$), a is the nanosphere radius, d is the centre-to-centre and h is the centre-to-interface distance

$$\mathbf{p}_n = \epsilon_1 \alpha(\omega) \left[\mathbf{E_0}(\mathbf{r}_n) + \sum_m \widehat{\mathbf{G}}(\mathbf{r}_n, \mathbf{r}_m)\, \mathbf{p}_m \right] \tag{6.3}$$

where m and n run from 1 to N, $\mathbf{r}_n$ is the position vector of the centre of the n-th sphere, $\mathbf{E_0}(\mathbf{r}_n)$ is the total external electric field (either the refracted one, if illuminated from the ε_2 side, or a superposition of the direct excitation and that reflected from the interface, if excited from the ε_1 side), and $\widehat{\mathbf{G}}(\mathbf{r}, \mathbf{r}')$ is Green's tensor of the total secondary dipole field:

$$\widehat{\mathbf{G}}(\mathbf{r}, \mathbf{r}') = \begin{cases} \widehat{\mathbf{G}}_\mathbf{0}(\mathbf{r}, \mathbf{r}') + \widehat{\mathbf{G}}_\mathbf{r}(\mathbf{r}, \mathbf{r}') & \mathbf{r} \neq \mathbf{r}' \\ \widehat{\mathbf{G}}_\mathbf{r}(\mathbf{r}, \mathbf{r}') & \mathbf{r} = \mathbf{r}' \end{cases} . \tag{6.4}$$

Here, $\widehat{\mathbf{G}}_\mathbf{0}(\mathbf{r}, \mathbf{r}')$ is Green's tensor in a homogeneous medium giving the electric field created at point $\mathbf{r}$ by a unit dipole located at $\mathbf{r}'$ [29]:

$$\widehat{\mathbf{G}}_\mathbf{0}(\mathbf{r}, \mathbf{r}') = \frac{1}{\epsilon_1} \left(\nabla\nabla + k_1^2 \mathbf{1} \right) \frac{\exp(i\, k_1 \left| \mathbf{r} - \mathbf{r}' \right|)}{\left| \mathbf{r} - \mathbf{r}' \right|} , \tag{6.5}$$

where $\nabla\nabla$ is the dyadic product of the nabla operators and $\mathbf{1}$ is the unit dyadic. $\widehat{\mathbf{G}}_\mathbf{r}(\mathbf{r}, \mathbf{r}')$ is Green's tensor of the reflected dipole field, which describes the interaction between nanoparticles mediated by the interface. This tensor can be calculated numerically using the Sommerfeld integrals formalism, as explained in Appendix B. The self-interaction term $\widehat{\mathbf{G}}_\mathbf{r}(\mathbf{r_n}, \mathbf{r_n})$ corrects the polarizability, Eq. (6.1), for the presence of the interface and guarantees, in particular, the correct energy balance.

For any given external field $\mathbf{E_0}(\mathbf{r})$, the system of equations in Eq. (6.3) is linear in $\mathbf{p}_n$, and the induced dipole moments can be computed by standard numerical methods. Once they are obtained, it is straightforward to calculate the total electric field at an arbitrary point $\mathbf{r}$ (located outside the volumes of the spheres):

$$\mathbf{E}(\mathbf{r}) = \mathbf{E_0}(\mathbf{r}) + \sum_m \widehat{\mathbf{G}}(\mathbf{r}, \mathbf{r}_m)\, \mathbf{p}_m \, . \tag{6.6}$$

In this work, the focus will be set in the radiation of the antenna (i.e., in the far-zone component of the scattered field) above the interface. Excitations from the ε_2 side (from below) under conditions of total reflection will be considered. Then, the incident field $\mathbf{E_0}(\mathbf{r})$ is an evanescent wave in the upper half-space, and its contribution to the total field is negligible if the detection point is sufficiently far from the interface. In this case, the far field is governed by the second term in Eq. (6.6). To obtain the antenna radiation pattern, one should calculate the angular dependence of the radiant intensity $U(\theta, \phi)$ on an enclosing sphere centred at the system and having a radius $R \gg \lambda_1$, where $\lambda_1 = 2\pi/k_1$ is the excitation wavelength in the medium. This intensity is given by

$$U(\theta, \phi) \propto R^2 \left| \sum_m \widehat{\mathbf{G}}(\mathbf{R}, \mathbf{r}_m)\, \mathbf{p}_m \right|^2 \, . \tag{6.7}$$

As shown below, radiation patterns of the antenna considered here can be highly directional: typically, they have a narrow main lobe and a set of much weaker side lobes. In order to characterize the directionality of the antenna radiation quantitatively, the standard directivity parameter will be used:

$$D = \frac{\max U(\theta, \phi)}{\int_{2\pi} U(\theta, \phi)\, d\Omega\, /2\pi} \, , \tag{6.8}$$

which is the ratio between the maximum radiant intensity and the average one. The integration in Eq. (6.8) is performed over the solid angle of 2π (instead of the usual 4π), because only the radiation in a half-space will be studied.

6.2 Numerical Results

The radiation patterns for an antenna comprising a linear array of $N = 15$ identical silver nanospheres were calculated. The array was in the proximity to an interface between two media with refractive indices $n_1 = \sqrt{\varepsilon_1} = 1.5$ (above the interface) and $n_2 = \sqrt{\varepsilon_2} = 2.1$ (below the interface). Other parameters of the model were chosen as follows: the nanosphere radius $a = 45$ nm, the inter-particle separation $d = 180$ nm, and the array-to-interface distance $h = 135$ nm. Tabulated data for the permittivity of silver [30] was used to calculate the polarizability of the nanospheres.

The system is illuminated by a plane wave with the wavelength in vacuum $\lambda_0 = 610$ nm, incident from the medium below the interface, and the radiation is detected in the upper half space. Let the spherical coordinates of the detection point be (R, θ, ϕ) with $\theta \in [0, \pi/2]$ and $\phi \in [0, 2\pi]$ ($\phi = 0$ in the direction of the

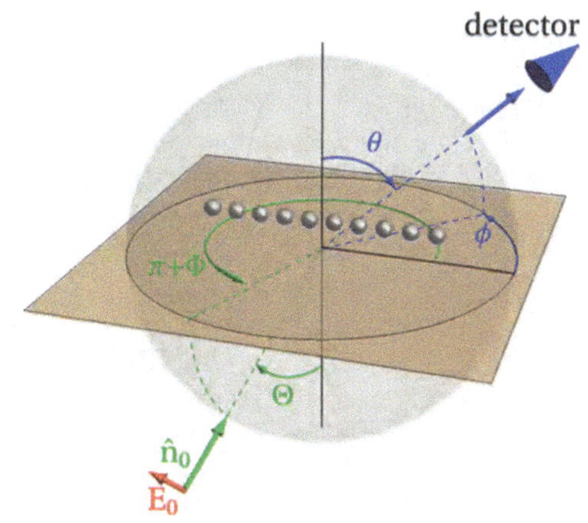

Fig. 6.2 The geometry of the excitation and detection: $\mathbf{E}_0$ is the excitation field, $\hat{\mathbf{n}}_0 = \mathbf{k}_2/k_2$ describes the direction of its incidence, $0 \leq \Theta \leq \pi/2$ and $-\pi/2 \leq \Phi \leq \pi/2$ are the polar and azimuth angles of incidence. The *blue cone* represents the detector whose angular coordinates are θ and ϕ

array), while Θ and Φ are the polar and azimuth angles of incidence of the incoming plane wave (see Fig. 6.2). To excite the system by an evanescent wave the incidence polar angle Θ should lie within an interval $(\Theta_c\,, \pi/2)$, where $\Theta_c = \arcsin(n_1/n_2)$ is the angle of the total reflection; for the considered system $\Theta_c = 45.6°$. Finally, the effect of the polarization of the excitation—*s* and *p*—will also be taken into account.

Figure 6.3 shows radiation patterns calculated for an s-polarized excitation incident at a polar angle $\Theta = 50° > \Theta_c$ and different azimuth angles: $\Phi = 0°$, 30°, 45°, 60°, 75°, 90°. The figure demonstrates that in all cases the radiation is highly directional; the directivity D is written in each plot. The interference between the fields scattered by the nanoparticles is constructive only within a relatively narrow solid angle, giving rise to the formation of the main lobe of the pattern. As shown below, its shape can be obtained analytically. It is clear from the comparison of the incidence geometry (see green arrows in Fig. 6.3) and the position of the main lobe that the latter is formed by the forward scattered light.

As can be seen from Fig. 6.3, the orientation of the main lobe is varying smoothly with the azimuth angle of incidence Φ. Much more pronounced and abrupt changes can occur if the incidence is more oblique. For example, for $\Theta = 60°$ the antenna main lobe orientation can be almost reversed by a relatively small change of the azimuth angle Φ, as shown in the lower panels of Fig. 6.4. In the corresponding upper panels, cross sections of the main lobes were plotted for the three different excitation conditions, which demonstrate that the predominant scattering switches from the forward to the backward one when the azimuth angle Φ changes from 30° to 10°. A similar switching effect is observed in the lower optically denser medium too (see Fig. 6.5). Note that the angles of the main radiation lobes are different from the angles of incidence and reflection; therefore, the excitation will not interfere with the signal, which can facilitate measurements.

The power radiated by a dipole in the proximity of a dielectric interface is higher in the optically denser medium. For the considered system about 10 % of the total scattered power goes into the upper half-space. The scattered intensity is proportional

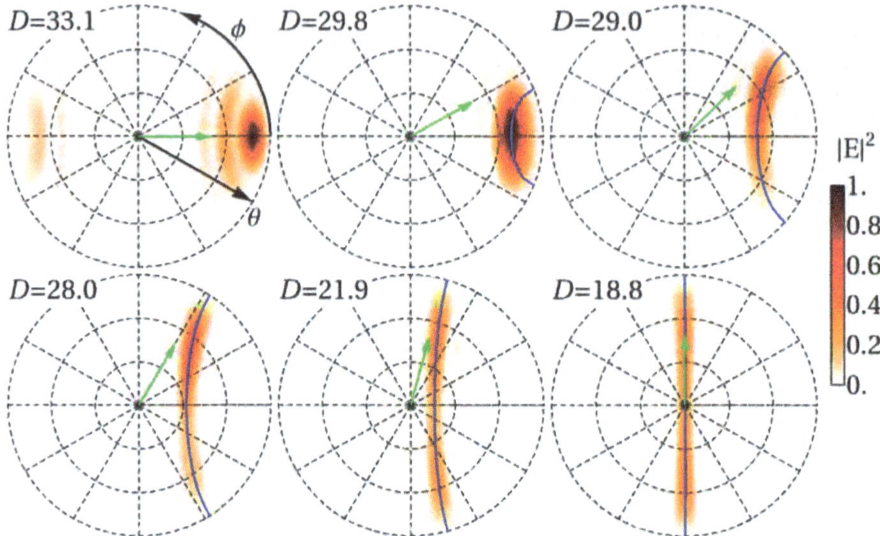

Fig. 6.3 Polar plots of the radiation patterns of the antenna described in the text. The *colour* gives the far-field intensity as a function of the detection angles θ and ϕ. The polar angle θ changes linearly along the radius from 0 to $\pi/2$ (*dashed circles* are drawn each 30°). The system was excited by an s-polarized plane wave with angles of incidence $\Theta = 50°$ and (from *left* to *right*) $\Phi = 0°,\ 30°,\ 45°$ (*upper row*), $60°,\ 75°,\ 90°$ (*lower row*). The excitation propagation direction is represented schematically by the *green vector*; its end point has the angle coordinates (Θ, Φ). The directivity D of the radiation pattern is given in the *upper left corner* of each plot. *Blue lines* represent the main lobe position calculated within the image dipole and stationary phase approximations, as described in the next section (see text for more details)

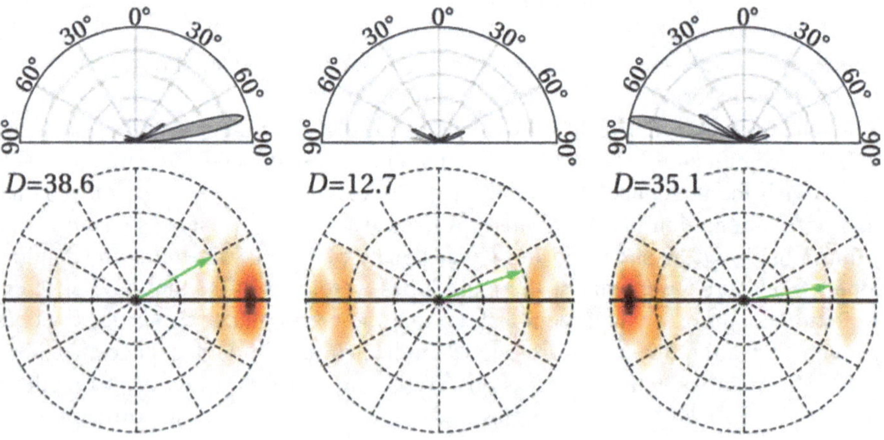

Fig. 6.4 *Lower row*—same as in Fig. 6.3, but for incidence at $\Theta = 60°$ and $\Phi = 30°,\ 20°,\ 10°$. Switching of the antenna radiation direction is observed when the azimuth angle Φ changes within a relatively narrow range. The *upper plots* show the cross sections of the radiation patterns computed along the *black thick lines* of the *lower plots*, i.e., $\phi = 0$, using *grey* filled areas. The *black solid lines* represent the same cross sections calculated according to the approximate formula Eq. (6.13)

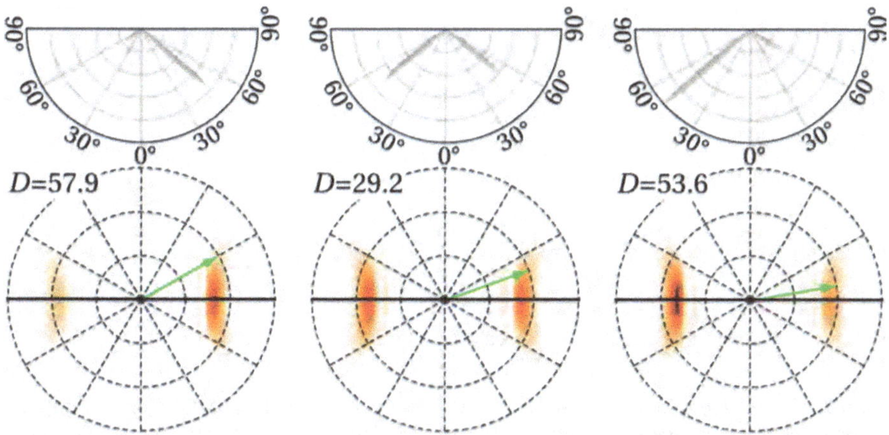

Fig. 6.5 Same as in Fig. 6.4, but for the lower half-space. The polar angle θ is measured from the normal to the interface pointing down

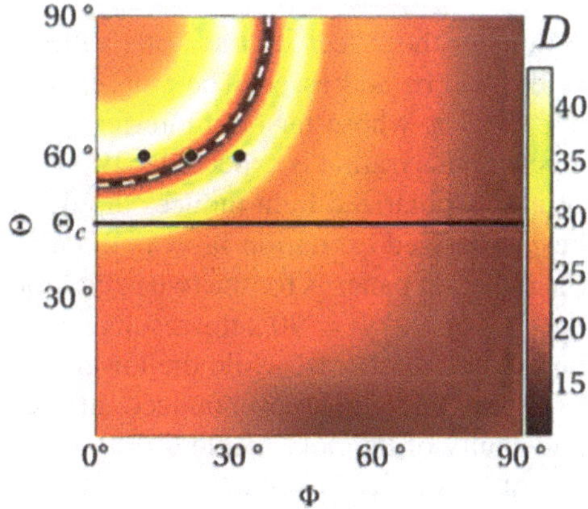

Fig. 6.6 The directivity versus the incidence angles Θ and Φ. The *dashed line* shows the incidence at which the phase difference between adjacent induced dipoles is π. The *dots* correspond to the illumination conditions used in Fig. 6.4

to the incident one and is limited therefore by the maximum possible dissipated power determined by the melting threshold of the MNPs.

Figure 6.6 shows the dependence of the directivity on the incidence angles Θ and Φ. The switching effect takes place in the vicinity of the pronounced dip in the directivity; the directivity is decreased when the phase difference induced by the external field between adjacent dipoles, $\mathbf{E}_0(\mathbf{r}_n)$ and $\mathbf{E}_0(\mathbf{r}_{n+1})$, equals π (see the white dashed line in the figure). In this case there is mirror symmetry in the far-field pattern with respect to the plane $\theta = \pi/2$ and two identical lobes of forward and backward scattered light co-exist. The two neighbouring light-coloured quarter-circle features are determined by the strong forward or backward scattering condition, when the far-field pattern is characterized by a single highly directional lobe. When the dip is crossed, the direction of the main lobe is switched abruptly.

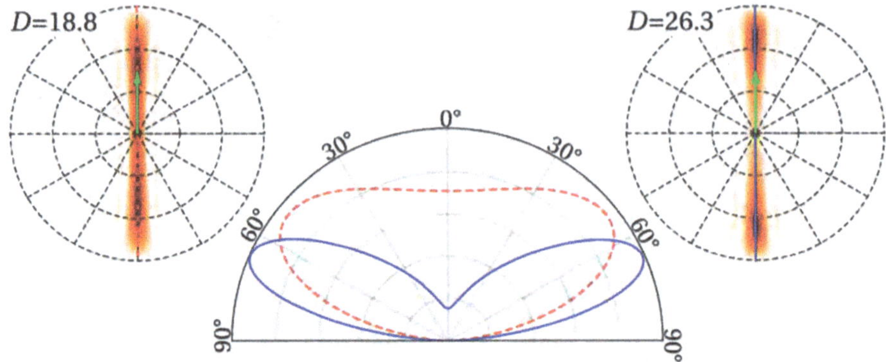

Fig. 6.7 *Upper plots*—radiation patterns for $\Theta = 46° \gtrsim \Theta_c$ and $\Phi = 90°$ for two different polarizations of the excitation: s (*left panel*) and p (*right panel*). The *lower, centred plot* shows cross sections of the two patterns at $\phi = \pi/2$: *red dashed* and *solid blue lines* correspond to s and p-polarized excitation, respectively

Finally, the effect of the polarization of the incoming light on the radiation pattern will be addressed. The polarization affects the direction of the induced particle dipoles, whose far field interference determines the shape of the pattern. The effect of polarization is expected to be most pronounced for $\Phi = 90°$, because in this case the s-polarized light induces dipoles almost parallel to the antenna axis, while the p-polarized excitation favours the dipole orientation perpendicular to the axis. The radiation patterns for the two excitation polarizations were calculated in the case of $\Theta = 46°$ and $\Phi = 90°$; the results are shown in Fig. 6.7. The upper panels present the full polar patterns, while the lower one shows their cross sections for $\phi = \pi/2$. As expected, the most pronounced difference between the two cases is observed in the vicinity of the polar angle $\theta = 0$ where the pattern has a pronounced dip in the case of p-polarized excitation while the radiation is strong in the other case.

Thus, the radiation pattern of the system can be adjusted by changing the excitation parameters, such as, the geometry of incidence or polarization, which can easily be controlled in experiment.

6.3 Analytical Results and Discussion

In this section an approximate analytical approach for the calculation of the antenna radiation pattern is presented, combining the image dipole and stationary phase approximations. Analytical results offer the opportunity of gaining insight into the physics of the system response and understanding qualitatively the numerical results obtained in the previous section. Moreover, the simple analytical formulas derived can be used to solve the inverse problem: engineer the geometry of the system and excitation meeting particular requirements, such as, a desired radiation pattern.

The radiation pattern is determined by the intensity of the secondary scattered field in the far zone. If a *single* dipole $\mathbf{p}$ is located at $(0, 0, h)$ (for convenience, Cartesian coordinates are used for dipole positions) and $R \gg h$, its radiation field reads

$$\mathbf{E}_{\mathrm{d}}(\theta, \phi) \approx \frac{k_1^2 \left[\mathbf{p} - (\mathbf{p} \cdot \mathbf{n})\, \mathbf{n}\right]}{\epsilon_1 R} e^{-ik_1 h \cos\theta} e^{ik_1 R} , \tag{6.9}$$

where $\mathbf{n}$ is the unit vector, pointing in the direction of the detection (θ and ϕ are its angular coordinates). The last exponential factor, determined by the overall phase $k_1 R$, is an angle-independent common factor and therefore it is irrelevant for the field intensity. Being in close proximity to the interface, the dipole induces surface charge density whose electric field contributes to the radiation as well. This secondary reflected field $\mathbf{E}_{\mathrm{r}}$ is given by the Sommerfeld integrals which can be calculated in the far-zone of the upper medium by making use of the stationary phase approximation [31]. The result reads:

$$\mathbf{E}_{\mathrm{r}}(\theta, \phi) = e^{2ik_1 h \cos\theta} \left[\mathcal{F}_s(\theta)\ (\mathbf{E}_{\mathrm{d}}(\theta, \phi) \cdot \mathbf{e}_s)\, \mathbf{e}_s + \mathcal{F}_p(\theta)\ \left(\mathbf{E}_{\mathrm{d}}(\theta, \phi) \cdot \mathbf{e}_p\right) \mathbf{e}_p \right] . \tag{6.10}$$

It this expression, $\mathbf{e}_s$ and $\mathbf{e}_p$ are the normalized vectors of the s- and p-polarized components of the radiated field,

$$\mathbf{e}_s = \frac{\mathbf{n} \times \mathbf{n}_z}{|\mathbf{n} \times \mathbf{n}_z|} \qquad \mathbf{e}_p = \mathbf{n} \times \mathbf{e}_s , \tag{6.11}$$

$\mathbf{n}_z$ being the normal to the interface pointing up, and $\mathcal{F}_s(\theta)$ and $\mathcal{F}_p(\theta)$ are the Fresnel reflection coefficients for the s- and p-polarized waves incident from the upper medium onto the interface [2],

$$\mathcal{F}_s(\theta) = \frac{n_1 \cos\theta - n_2 \cos\theta_t}{n_1 \cos\theta + n_2 \cos\theta_t} \qquad \mathcal{F}_p(\theta) = \frac{n_2 \cos\theta - n_1 \cos\theta_t}{n_2 \cos\theta + n_1 \cos\theta_t} , \tag{6.12}$$

with θ_t given by Snell's law, $n_1 \sin(\theta) = n_2 \sin(\theta_t)$. As seen from the comparison of Eq. (6.9) and Eq. (6.10), the secondary field $\mathbf{E}_{\mathrm{r}}$ is equivalent to that produced by an image dipole placed at $(0, 0, -h)$ in a homogeneous medium with the dielectric constant ϵ_1. However, the s and p components of the reflected field $\mathbf{E}_{\mathrm{r}}$ are renormalized by the corresponding Fresnel reflection coefficients.

The radiation of an array of dipoles is the superposition of their far fields; its intensity angular distribution depends exclusively on the relative phases of the contributing fields (a relatively small shift of the coordinate origin would result in an additional common angular independent, and therefore irrelevant, phase of the field). When such dipoles are arranged in a regular linear chain and $R \gg Nd$, the total secondary far field can be calculated analytically, using the following additional approximations. First, the phase and orientation of the n-th dipole are assumed to be determined by the phase and polarization of the external field $\mathbf{E}_0(\mathbf{r}_n)$ acting upon it and, second, that all dipoles have the same amplitude. It turns out that for the considered system, excitation geometries and wavelengths these approximations are reasonable.[1] In this case, the phase difference between two dipoles located at $(x_1, 0, h)$ and $(x_2, 0, h)$ is $k_1 n_{21}(x_2 - x_1) \sin\Theta \cos\Phi$ with $n_{21} = n_2/n_1$. On the other hand, the phase difference in the far fields of the two dipoles is $k_1(x_2 - x_1) \sin\theta \cos\phi$. Using all the above assumptions and the corresponding phase relations, one can sum the fields produced by all dipoles and their images, arriving finally at the following expression for the radiation[2]:

$$\mathbf{E}(\theta, \phi, \Theta, \Phi) = \frac{\sin(\beta N/2)}{\sin(\beta/2)} \left[\mathbf{E}_{\mathrm{d}}(\theta, \phi) + \mathbf{E}_{\mathrm{r}}(\theta, \phi) \right], \tag{6.13}$$

where β is the phase difference of the radiation of two neighboring particles,

$$\beta = k_1 d \left(\sin\theta \cos\phi - n_{21} \sin\Theta \cos\Phi\right). \tag{6.14}$$

In the adopted approximation, the orientation of the induced dipoles is dictated by the polarization of the incident field $\mathbf{E}_0$. If the excitation is s-polarized, the dipoles are oriented along $\Phi + \pi/2$ and no field is radiated in this direction. If the detection angle ϕ is not close to $\Phi + \pi/2$, the fields in the square brackets in Eq. (6.13) are smooth functions of θ and ϕ. Therefore, it is the fraction of the two sine functions in Eq. (6.13) that determines the lobe structure of the radiation pattern (a detailed discussion of this pre-factor can be found, for example, in [32]). This fraction is large when $\beta = 2\pi n$ with an integer n, in which case the interference is constructive and the far-field of N dipoles is about N times larger than that of a single dipole. Hence, the enhancement factor of the radiated intensity over that of a single nanoparticle is on the order of N^2. Strong far-field lobes are formed in the vicinity $\delta\beta \ll 2/N$ of the scattering angles giving solutions to the equation $\beta = 2\pi n$. This vicinity becomes smaller as N increases, resulting in the narrowing of the lobes. On the contrary, if θ and ϕ are such that $\beta N = 2\pi n$ where n is not a multiple of N, the interference is destructive, and the antenna is only weakly radiating in these directions. Essentially, Eq. (6.13) gives the far-field of a double chain of identical dipoles and their images.

The upper panels of Fig. 6.4 show the radiation pattern cross-sections calculated numerically, according to Eq. (6.7), (solid grey regions) and those obtained using the approximate analytical expression in Eq. (6.13) (solid black lines). The amplitude of the analytical result was scaled to get the same maximum as the numerical result. The figure demonstrates that the analytical expression gives an excellent description of the main lobes of the patterns. For the chosen polar angle of incidence $\Theta = 60°$, the main lobe almost reverses its direction when the azimuth angle of incidence changes from $\Phi = 10°$ to $\Phi = 30°$. Below, a simple qualitative explanation of this behaviour is provided, based on the analysis of our approximate formula.

---

[1] These approximations are exact for an infinite system.

[2] The derivation of the formula is analogous to the one made by Von Laue to study the X-ray diffraction in crystalline structures.

To understand the underlying physics of the predominant scattering direction switching, it is useful to rewrite the phase β in terms of the projection of the incoming and outgoing wave vectors onto the chain axis: $\beta = (k_{1x} - k_{2x})\, d$. The interference is constructive when $\beta = 2\pi n$, which can be interpreted as the refraction condition: $k_{1x} - k_{2x} = 2\pi n/d$, where $2\pi n/d$ are the vectors of the reciprocal lattice with the lattice constant d. The latter relationship is the momentum conservation law for a periodic structure where the k-vector is conserved to within a vector of the reciprocal lattice.

In the system studied, the fact that the lower media is optically denser than the upper one (i.e., $k_2 > k_1$) plays the key role and results in the appearance of qualitatively new solutions as compared to the traditional refraction in a homogeneous environment. As long as $k_{2x} \leq k_1$ the condition $\beta = 0$ can be met for $\theta \approx \Theta$ and $\phi \approx \Phi$, i.e., the direction of the main lobe is close to that of the incidence resulting in strong "forward" scattering. Moreover, for the range of parameters chosen, $k_1 d < \pi$, and therefore the scattering pattern is always characterized by a single strong lobe.

On the other hand, for $k_{2x} > k_1$, forward scattering is forbidden because it would require a wave vector larger than the one allowed in the upper medium. In this qualitatively new situation, the constructive interference occurs for the outgoing wave vectors which differ from the incoming ones by a vector of the reciprocal lattice. In particular, the condition $\beta = -2\pi$ can be met when k_{1x} and k_{2x} have opposite signs. The latter is the "backward" scattering observed in Fig. 6.4.

Figure 6.7 demonstrates that the antenna radiation patterns can be controlled also by the polarization of the incident light, which is a parameter that can easily be changed in an experiment. In Fig. 6.7, the polar angle of incidence Θ is only slightly larger than the angle of total reflection while the azimuth angle $\Phi = \pi/2$ implies incidence normal to the chain axis. Under these conditions, the main lobe of the pattern is determined by the equation $\beta = 0$, having the obvious solution $\phi = \pi/2$. The θ-dependence of the patterns in this case is dictated by the excitation polarization. For the s-polarized excitation, all dipoles are oriented along the chain, that is, along $\phi = 0$, and the radiation is efficient within a continuous broad range of polar angles θ (see the dashed red line in the lower panel of Fig. 6.7). Contrary to that, when the excitation is p-polarized, all dipoles are almost perpendicular to the chain axis and to the interface plane; that is, along $\theta = 0$. According to Eq. (6.9), a dipole does not radiate in its own direction, which explains the appearance of a pronounced dip around $\theta = 0$ in the radiation pattern (see the blue solid line in the lower panel of Fig. 6.7).

6.4 Summary

The radiation patterns of a plasmonic antenna were studied. The antenna comprised a regular linear chain of identical metal nanospheres in close proximity to an interface between two media with high dielectric contrast. When such a system is illuminated from the higher refractive index side of the interface it can be excited by evanescent waves. In this case the excitation does not mask the useful antenna signal, which is advantageous for measurements and applications. It was proved that the radiation pattern and its directivity can be controlled by changing the incidence angles and/or polarization of the excitation. In particular, for some excitation geometries, the antenna pattern is characterized by a very narrow main lobe whose direction can be changed abruptly by a relatively small change of incidence angles.

Initially, the antenna radiation patterns were calculated using the traditional, much more numerically elaborated and accurate, Sommerfeld integrals approach. Then a much simpler one, based on the image dipole and the stationary phase approximations was proposed. It was shown that, despite the complexity of the system, the analytical expression for the radiation patterns gives an excellent description of the main features of the antenna response. These simple formulas can become a useful tool for solving the inverse problem: engineering a system that has a desired radiation pattern.

An antenna comprising identical MNPs operates within a relatively narrow band width; however, the spectral range can be broadened by using graded plasmonic arrays [14, 15]. It should be noted that, although the theoretically simplest case of the chain of nanospheres was considered, the results are expected to be valid for arrays of particles with more complex shapes, such as discs, which can be easier to fabricate.

Similar ideas and approaches can be applied to more complicated nanostructures, such as 2D arrays or metamaterials [33]. The dielectric interface would still play its key role, contributing two important aspects: the dielectric contrast is an additional degree of freedom allowing radiation patterns of the nanoscopic sources of light to be controlled, while excitation by evanescent waves results in the conversion of a macroscopic plane wave into a narrow beam of light with adjustable characteristics and direction. Apart from being an interesting fundamental phenomenon, this opens new possibilities in optical nano-devices design and new opportunities to control the flow of electromagnetic energy at the nanometre scale, in particular, for precisely addressing and exciting nanoscopic objects such as nanostructures, quantum dots, single molecules, etc.

References

1. S.A. Maier, H.A. Atwater, Plasmonics: localization and guiding of electromagnetic energy in metal/dielectric structures. J. Appl. Phys. **98**, 011101 (2005)
2. L. Novotny, B. Hecht, *Principles of Nano-Optics* (Cambridge University Press, Cambridge, 2006)
3. S. Maier, *Plasmonics: Fundamentals and Applications*. (Springer, New York, 2007)
4. E. Ozbay, Plasmonics: merging photonics and electronics at nanoscale dimensions. Science **311**, 189–193 (2006)
5. R. de Waele et al., Tunable nanoscale localization of energy on plasmon particle arrays. Nano Lett. **7**, 2004–2008 (2007)
6. H.A. Atwater, A. Polman, Plasmonics for improved photovoltaic devices. Nat. Mater. **9**, 205–213 (2010)

7. M.L. Brongersma, Plasmonics: engineering optical nanoantennas. Nat. Photonics **2**, 270–272 (2008)
8. P. Bharadwaj et al., Optical antennas. Adv. Opt. Photonics **1**, 438–483 (2009)
9. L. Novotny, N. van Hulst, Antennas for light. Nat. Photonics **5**, 83–90 (2011)
10. E.S. Barnard et al., Spectral properties of plasmonic resonator antennas. Opt. Express **16**, 16529–16537 (2008)
11. Z. Zhang et al., Manipulating nanoscale light fields with the asymmetric Bowtie nanocolorsorter. Nano Lett. **9**, 4505–4509 (2009)
12. Y.-Y. Yang et al., Steering the optical response with asymmetric bowtie 2-color controllers in the visible and near infrared range. Opt. Commun. **284**, 3474–3478 (2011)
13. T. Coenen et al., Directional emission from plasmonic Yagi-Uda antennas probed by angle-resolved cathodoluminescence spectroscopy. Nano Lett. **11**, 3779–3784 (2011)
14. A.V. Malyshev et al., Frequency-controlled localization of optical signals in graded plasmonic chains. Nano Lett. **8**, 2369–2372 (2008)
15. R.S. Pavlov et al., Log-periodic optical antennas with broadband directivity. Opt. Commun. **285**, 3334–3340 (2012)
16. P. Biagioni et al., Cross resonant optical antenna. Phys. Rev. Lett. **102**, 256801 (2009)
17. R. Bardhan et al., Metallic nanoshells with semiconductor cores: optical characteristics modified by core medium properties. ACS Nano **4**, 6169–6179 (2010)
18. T.H. Taminiau et al., Optical nanorod antennas modeled as cavities for dipolar emitters: evolution of sub- and super-radiant modes. Nano Lett. **11**, 1020–1024 (2011)
19. S. Palomba et al., Nonlinear plasmonics with gold nanoparticle antennas. J. Opt. A: Pure Appl. Opt. **11**, 114030 (2009)
20. N. Liu et al., Nanoantenna-enhanced gas sensing in a single tailored nanofocus. Nat. Mater. **10**, 631–636 (2011)
21. P. Bharadwaj et al., Nanoscale spectroscopy with optical antennas. Chem. Sci. **2**, 136–140 (2011)
22. S.Y. Park, D. Stroud, Surface-plasmon dispersion relations in chains of metallic nanoparticles: an exact quasistatic calculation. Phys. Rev. B **69**, 125418 (2004)
23. J.-Y. Yan et al., Optical properties of coupled metal-semiconductor and metalmolecule nanocrystal complexes: role of multipole effects, Phys. Rev. B **77**, 165301 (2008)
24. F.J. García de Abajo, A. Howie, Retarded field calculation of electron energy loss in inhomogeneous dielectrics. Phys. Rev. B **65**, 115418 (2002)
25. J.M. Gérardy, M. Ausloos, Absorption spectrum of clusters of spheres from the general solution of Maxwell's equations. II. optical properties of aggregated metal spheres. Phys. Rev. B **25**, 4204–4229 (1982)
26. M. Meier, A. Wokaun, Enhanced fields on large metal particles: dynamic depolarization. Opt. Lett. **8**, 581–583 (1983)
27. G. Mie, Beiträge zur Optik trüber Medien, speziell kolloidaler Metallosungen. Annalen der Physik **25**, 377–445 (1908)
28. C.F. Bohren, D.R. Hugffman, *Absorption and Scattering of Light by Small Particles* (Wiley, New York, 1983)
29. M.S. Tomaš, Green function for multilayers: light scattering in planar cavities. Phys. Rev. A **51**, 2545–2559 (1995)
30. E. Palik, *Handbook of Optical Constants of Solids*. (Academic Pres, Boston, 1998)
31. W. Chew, *Waves and Fields in Inhomogeneous Media*. (IEEE Press, New York, 1999)
32. C. Balanis, *Antenna Theory: Analysis and Design*. (Wiley, New York, 1982)
33. E. Prodan et al., A hybridization model for the plasmon response of complex nanostructures. Science **302**, 419–422 (2003)

Chapter 7
Electro-Optical Hysteresis of Nanoscale Hybrid Systems

Semiconductor Quantum Dots have many applications in optics due to their atomic-like spectra and have long been pledged to be an essential constituent of future quantum logical systems [1]. The latter is due to their most celebrated characteristics, such as long decoherence times reaching tenths of nanoseconds [2], spin decoherence times in the millisecond range [3], or typical sizes below 10 nm. The shift of the energy levels (discrete due to quantum confinement effects) under the effect of electrostatic gates has allowed single-electron transport through arrays of SQDs to be experimentally tested in numerous occasions [2, 4]. Nowadays, it is possible to engineer their energy levels by varying their shape and size [5–7]. Typically, nanocrystals with sizes in the range 1–10 nm have allowed transitions laying in the optical range, which paves the way for the design of ultrafast quantum memories [8–10].

In the proximity of reflecting systems, such as MNPs or interfaces, irradiated SQDs can exhibit strong nonlinear effects, which can be used to design electronic components. In the simplest quasi-static approximation such systems can be considered as an artificial hybrid quasi-molecule: when it is optically excited, the dipole moment of the optical transition in the SQD generates image dipoles whose electric field acts back upon the real dipole, providing an electromagnetic feedback. This mechanism has long been studied for SQD-MNP systems: the presence of a "resonator", the MNP in the latter case, also leads to self action (feedback) of the SQD. Together with the nonlinearity of the SQD itself, this can give rise to a variety of interesting optical properties [11–15]. In particular, if the coupling between two nanoparticles is strong enough, the self-action can result in hysteresis in the optical response [16].

In this chapter, a simple hybrid system comprising a SQD placed close to an interface between two materials is considered. It is shown that the SQD-interface quasi-dimer, which seems to be simpler to fabricate than the previously considered SQD-MNP complexes, can manifest nonlinear optical properties. If one of the materials forming the interface allows for the control of the electron concentration, and therefore of the local refractive index, the optical bistability and hysteresis manifesting itself at nanoscale can be controlled macroscopically, e.g., by changing

J. Munárriz Arrieta, *Modelling of Plasmonic and Graphene Nanodevices*,
Springer Theses, DOI: 10.1007/978-3-319-07088-9_7,

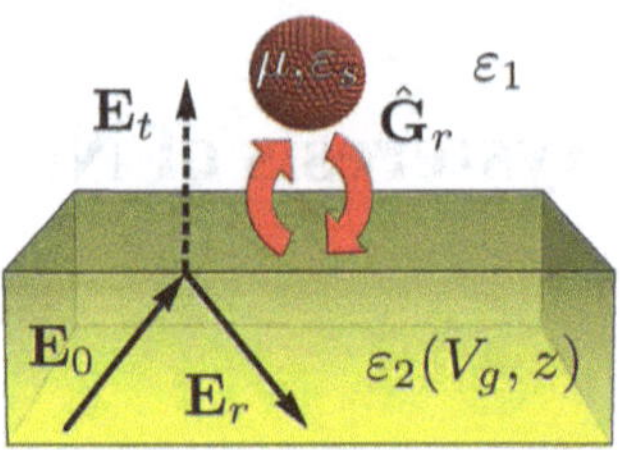

Fig. 7.1 Schematics of the hybrid SQD-interface system, the SQD is embedded into a homogeneous dielectric background with permittivity ε_1 and subjected to an external field $\mathbf{E}_t$. $\boldsymbol{\mu}$ and ε_s denote the SQD optical transition dipole moment and the semiconductor dielectric constant, respectively, while $\varepsilon_2(V_g, z)$ is the tunable gradient corresponding to the dielectric permittivity of the substrate. The *curved arrows* symbolize the dipole self-interaction of the optical transition dipole of the SQD, governed by Green's tensor of the reflected field $\widehat{\mathbf{G}}_r$

the gate voltage. Such a control of the refractive index has been demonstrated in an indium tin oxide (ITO)—SiO_2 heterostructure [17]. Below, it is shown that in the case of the latter interface the system can manifest not only optical but also electro-optical bistability and hysteresis. The fact that the inorganic systems can sustain high electric fields suggests such possible applications of the hybrid system as electro-optical switches, modulators or memory cells at the nanoscale in the visible spectrum.

7.1 Approximations and Formalism

A SQD embedded in a dielectric host grown on top of a substrate material is considered. Its centre is set to be at $\mathbf{r} = (0, 0, z)$. The host and the substrate are characterized by their dielectric permittivities ε_1 and $\varepsilon_2(V_g, z)$, respectively (see Fig. 7.1). The system is driven by an incident electric field with direction of incidence $\hat{\mathbf{k}}_0$, amplitude $\mathbf{E}_0$ and frequency ω_0.

We consider a CdSe (or CdSe/ZnSe) SQD, widely used in the literature [6, 18–20]. If the detuning between the driving and the transition frequency ω, $\Delta \equiv \omega_0 - \omega$, is small compared to the separation between excited states, the SQD can be treated as a two-level system. Its interaction with an external field is governed by the dipole moment $\boldsymbol{\mu}$ of the optical transition between the ground and the first excited state.

The spontaneous decay and dephasing are key elements which govern the dynamics of these systems, and need to be included in any realistic model. Here, this is done by means of their corresponding relaxation constants, γ (decay) and Γ (dephasing). As a consequence, the time evolution of the two-level system alone is no longer unitary, and its state needs to be defined statistically, using the density matrix formalism,

$$\hat{\rho} = \begin{pmatrix} \rho_{ee} & \rho_{eg} \\ \rho_{ge} & \rho_{gg} \end{pmatrix} . \tag{7.1}$$

Within the semiclassical formulation, the electric field is considered classically, and the dynamics of the system is governed by the optical Bloch equations [21], which read:

$$\dot{Z} = -\gamma(Z+1) - \frac{1}{2}\left[\Omega P^* + \Omega^* P\right]$$
$$\dot{P} = -(i\Delta + \Gamma)P + \Omega Z\,. \tag{7.2}$$

Here, Z and P are the population difference between the excited and ground state and the amplitude of the off-diagonal density matrix element, respectively:

$$Z = \rho_{ee} - \rho_{gg} \qquad\qquad P = 2i\rho_{eg}\exp(i\omega t)\,, \tag{7.3}$$

and $\Omega = \boldsymbol{\mu}\mathbf{E}/\hbar$, with $\mathbf{E}$ being the electric field acting *inside* the SQD.

Using the density matrix formalism, $\mathbf{P}_{\mathrm{SQD}} = \langle\mathbf{p}\rangle = \mathrm{Tr}(\hat{\boldsymbol{\rho}}\,\mathbf{p}) = -i\boldsymbol{\mu}P$. Due to its size, the action of the SQD optical dipole moment $\mathbf{P}_{\mathrm{SQD}}$ upon the rest of the system is considered within the point dipole approximation. Nevertheless, when treating the electric field inside the SQD and its electric polarization, it is important to take into account the finite size of the dot. The exciton radius a_B in CdSe is about 5 nm [5], so the wave functions involved in the optical transition are extended over the whole dot. Therefore, the approximation of a homogeneous electric polarization of the whole SQD volume by the optical transition dipole moment $\mathbf{P}_{\mathrm{SQD}}$ can be used. Finally, the screening needs to be taken into account, which is done including the factor $\varepsilon_s' = (\varepsilon_s + 2\varepsilon_1)/(3\varepsilon_1)$, where ε_s is the dielectric permittivity of the SQD [22]. These considerations yield the following expression for the electric field *inside* the quantum dot:

$$\mathbf{E} = \frac{1}{\varepsilon_s'}\left[\mathbf{E}_{0,\mathrm{in}}(\mathbf{r}) + \widehat{\mathbf{G}}_r(\mathbf{r},\mathbf{r})\,\mathbf{P}_{\mathrm{SQD}}\right]\,, \tag{7.4}$$

where $\mathbf{E}_{0,\mathrm{in}}(\mathbf{r})$ is the field at $\mathbf{r}$ due to the external excitation, and $\widehat{\mathbf{G}}_r(\mathbf{r},\mathbf{r})\,\mathbf{P}_{\mathrm{SQD}}$, the field due to the reflection of the dipole moment $\mathbf{P}_{\mathrm{SQD}}$ in the substrate, written in terms of reflected Green's tensor $\widehat{\mathbf{G}}_r(\mathbf{r},\mathbf{r})$ (for further details, see Appendix B.2).

The field $\mathbf{E}_{0,\mathrm{in}}(\mathbf{r})$ due to the incoming plane wave is calculated using the transmitted plane wave, if the excitation comes from the lower medium, or summing up the incident and the reflected ones, for waves coming from above. Both terms can be calculated using the TMM detailed in Appendix B.1, and written in compact form as

$$\mathbf{E}_{0,\mathrm{in}}(\mathbf{r}) = \widehat{\mathbf{T}}(\mathbf{r})\cdot\mathbf{E}_0\,,$$

where $\widehat{\mathbf{T}}(\mathbf{r})$ is the tensor that returns the field at $\mathbf{r}$ due to the incoming plane wave.

Therefore, the Rabi frequency Ω which enters the optical Bloch equations in Eq. (7.2) can be represented in the following form:

$$\Omega = \widetilde{\Omega}_0 - iGP, \tag{7.5}$$

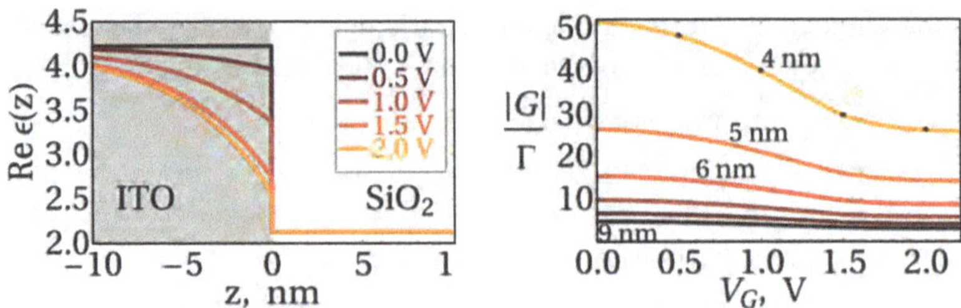

Fig. 7.2 *Left* variation of the dielectric permittivity at an SiO_2-ITO interface as a function of the distance z and applied gate voltage, V_g, for $\lambda = 525\,\text{nm}$. *Right* behaviour of the feedback parameter G as a function of the gate voltage is shown, with the SQD placed at different heights above the interface. The rest of parameters are given in the text

with $\widetilde{\Omega}_0$ and G given by

$$\widetilde{\Omega}_0 = \frac{\boldsymbol{\mu} \cdot \widehat{\mathbf{T}}(\mathbf{r}) \cdot \mathbf{E}_0}{\varepsilon_s' \hbar}, \qquad G = \frac{\boldsymbol{\mu} \cdot \widehat{\mathbf{G}}_r(\mathbf{r}, \mathbf{r}) \cdot \boldsymbol{\mu}}{\varepsilon_s' \hbar}, \tag{7.6}$$

where G is the feedback parameter.

Here, a SiO_2–ITO interface is considered, with a dielectric constant of the host $\varepsilon_1 = 2.16$. The dielectric permittivity for the substrate $\epsilon_2(V_g, z)$ has been obtained from tabulated data [17], as shown in the left panel of Fig. 7.2. It is important to stress that the change in the permittivity close to the interface is due to the creation of an accumulation layer with depth $\sim$5 nm and carrier concentrations as high as $2.8 \times 10^{22}\,\text{cm}^{-3}$. The small depth, compared to the wave length, produces only a very small change in the reflection of plane waves at the interface, but it strongly affects the near field, leading to a large shift in the feedback parameter, as shown in the right panel of Fig. 7.2.

In the calculations, exact Green's tensor of the reflected dipole field is determined using Sommerfeld integrals containing Fresnel coefficients appropriate for the studied heterostructure. For the case of a stratified media, these integrals can be calculated following [23], substituting the continuously varying dielectric gradient $\epsilon_2(V_g, z)$ by an appropriate number of layers with constant dielectric permittivity such that $G(V_g)$ converges to a constant value as the number of layers increases. The same result can be approximated by using the multiple-scattering integral formalism presented in [24]. The convergence for the parameters used in this work is shown in Fig. 7.3.

The effect of the feedback on the dynamics of the system governed by the optical Bloch equations with such self-action has been discussed in detail in [11–16, 25, 26]. It is well known that it plays a key role in many nonlinear effects predicted for hybrid systems. For example, the feedback should exceed a certain threshold for the optical bistability to occur [25, 27]. Note also that the the feedback should be compared to the damping rate Γ which increases with temperature [28], so the larger is the value of G the higher the temperature at which the predicted effects would be observed.

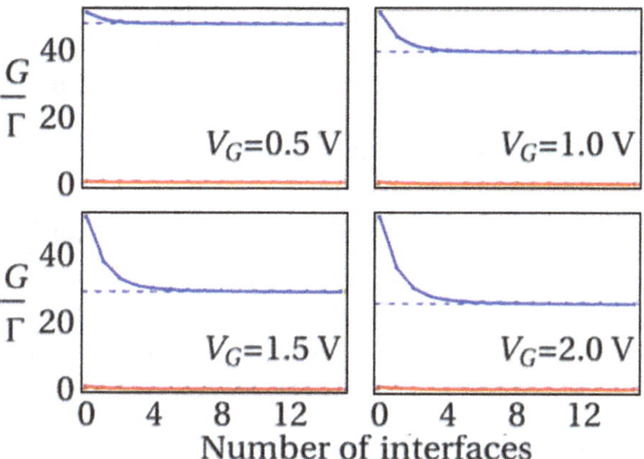

Fig. 7.3 Convergence of Re(G)—*blue*—and Im(G)—*red*—with the number of layers used to simulate the dielectric permittivity profile of the substrate, for the 4 *black points* marked in the *right panel* of Fig. 7.2. *Dashed horizontal lines* represent a continuous approximation, calculated using [24]

7.2 Bistability Due to Self-Interaction

A CdSe/ZnSe SQD in the proximity of a SiO_2-ITO interface, excited by a plane wave with $\lambda = 525$ nm, is considered. The plane wave propagates from the lower to the upper medium forming an angle of 45° with the interface, and is polarized parallel to it (s polarization). The following parameter set will be used: the transition energy $\hbar\omega_0 = 2.36$ eV (corresponding to the optical transition in a 3.3 nm SQD), the SQD dielectric constant $\varepsilon_s = 6.2$, the SQD centre-to-interface distance $z = 6$ nm, the SQD transition dipole moment $\mu = 0.65$ eV $\cdot$ nm [11] and the relaxation constants $1/\gamma = 0.8$ ns and $1/\Gamma = 0.3$ ns [12].

Figure 7.4 shows the population difference Z and the polarization intensity $|\mathbf{P}|^2$ of the stationary state ($\dot{Z} = \dot{P} = 0$, see [16]) versus the excitation intensity for the set of parameters specified above, $V_g = 0$V and different detunings from resonance Δ. The excitation intensity is given in terms of the *bare* Raby frequency,

$$\Omega_0 = \frac{|\mu||\mathbf{E}_0|}{\hbar} ,$$

where $\mathbf{E}_0$ is the field at $z = 0$, if the space were homogeneous, with $\epsilon = \epsilon_2(V_g = 0, z)$.

Figure 7.5 shows the regions of the parameter space $\{V_g,\ \Delta,\ z\}$ which manifest optical bistability. This means that, for a range of excitation intensities Ω_0, they present two stable stationary solutions. The figure shows that the bistability is a common feature of the simulated system for a wide range of parameters, as long as the feedback parameter G is large enough.

For $V_g = 0$ V, the field dependences of Z and P, have three allowed values for a given intensity $|\Omega_0|^2/(\gamma\Gamma)$ within a window $-1.7\,\Gamma \leq \Delta \leq 12.9\,\Gamma$. This bistability range shrinks to $-1.1\,\Gamma \leq \Delta \leq 8.4\,\Gamma$ for $V_g = 1.3$ V (see the black dashed vertical lines in Fig. 7.5). From now on this parameter will be set to $\Delta = 0$, i.e., the system will be excited with the frequency given by the SQD resonance.

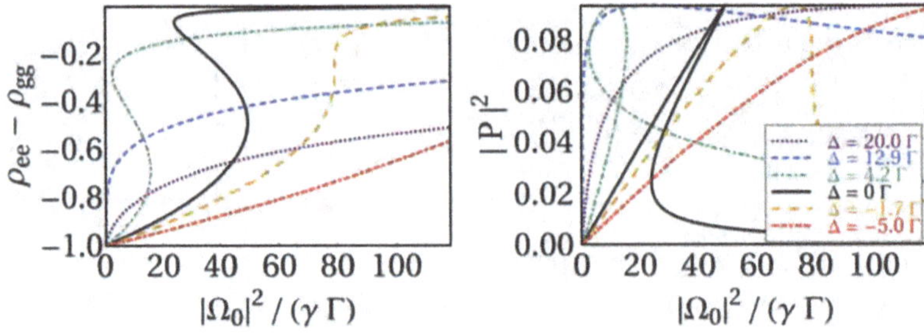

Fig. 7.4 Population difference $Z = \rho_{ee} - \rho_{gg}$ (*left*) and SQD polarization intensity $|P|^2$ (*right*) for different detunings Δ, as functions of the external field intensity $\Omega_0/\sqrt{\gamma\Gamma}$, with no gate voltage V_g

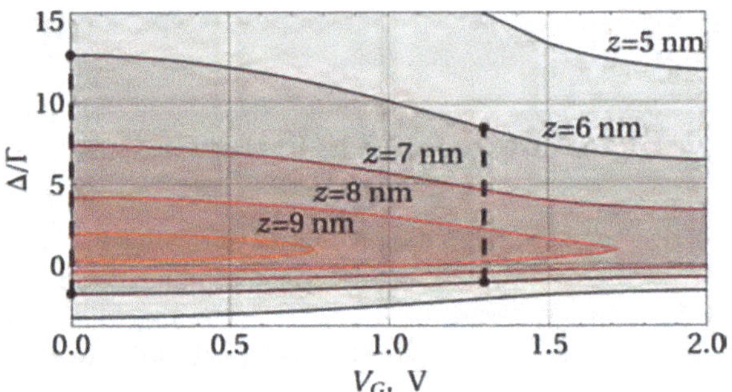

Fig. 7.5 Regions with optical bistability versus V_g, Δ, z. For each z, the region enclosed by the lines can sustain stable states with different populations and polarizations

7.3 Electro-Optical Hysteresis

Using the configuration described above, the stationary solutions were calculated for a wide range of values $\{V_g, \Omega_0\}$, as shown in Fig. 7.6, where the regions without bistability are marked in blue, whereas the bistable regions are plotted in red (stable branches) and grey (unstable one). If the parameters V_g, Ω_0 are adiabatically swept back and forth across the bistable region, the time evolution of Z and P, calculated integrating Eq. (7.2), shows hysteresis loops, with sharp changes at the boundaries of the bistable regions. This is addressed in Fig. 7.7 for the case of constant $V_g = 0\text{V}$, plotting both the stationary solutions and the behaviour of the dynamic simulation. The intensity of the incoming light is changed using a triangular profile, going from $|\Omega_0|^2 / (\gamma\Gamma) = 0$ to a maximum value of 60 and then returning to 0. The complementary time-domain simulation, with constant $|\Omega_0|^2 / (\gamma\Gamma) = 21$ and V_g being swept back and forth across the bistable region using the triangular profile, with the maximum voltage $V_g = 1.65\,\text{V}$, is shown in Fig. 7.8.

To observe the hysteresis, the initial and final points should be monostable, so that the system can be flipped in both directions. This is easily met when V_g is kept constant and Ω_0 sweeps back and forth, as shown in Fig. 7.6. It is possible to find full hysteresis loops for any fixed value of V_g, but the bistability region narrows as the gate voltage is raised. This is due to the fact that the self-interaction term G decreases,

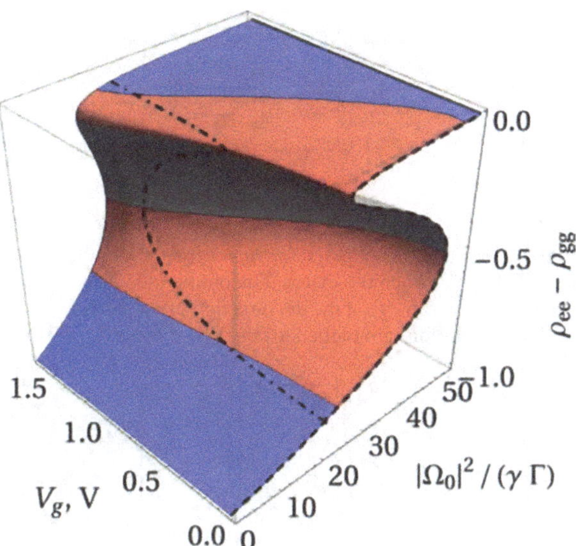

Fig. 7.6 Stationary solutions of the population difference Z in the system, as a function of the incoming field and the gate voltage, for $\Delta = 0\,\Gamma$, $z = 6\,\text{nm}$. The regions without bistability are marked in *blue*, whereas the bistable regions are plotted in *red* (stable branches) and *grey* (unstable one). For a constant voltage $V_g = 0$ (*black dashed line*) and a constant external incoming field $|\Omega_0|^2/(\gamma\Gamma) = 21$ (*dash-dotted line*), hysteresis loops are shown in Figs. 7.7 and 7.8, respectively

because the profile of the dielectric gradient becomes smoother. As a consequence, the layer turns more transparent to the evanescent waves, which dominate the self-interaction when $z \ll \lambda$. Both the change in the dielectric permittivity $\epsilon(z)$ and its effect on the self-interaction term are shown in Fig. 7.2.

Hysteresis loops controlled by the gate voltage require a careful choice of the excitation intensity Ω_0, as there is a narrow region for which hysteresis loops can be attained. Nevertheless, this region can be widened by placing the SQD closer to the interface. As the dipole and its image get closer, the fastest decaying evanescent waves dominate the interaction, and these waves are the most affected by changes of the local permittivity at the interface.

7.4 Non-Adiabatic Branch Flips

The speed at which the system can be flipped back and forth is among its most prominent features, both for its fundamental interest and its possible integration in devices operating at high frequencies. It can be tested by setting a constant illumination where electro-optical hysteresis is known to occur, e.g., $|\Omega_0|^2/(\gamma\Gamma) = 22$. Then, and initial voltage is chosen approximately in the middle of the bistable region, $V_{g,0} = 1.1\,\text{V}$. For this parameters, the system shows bistability. Then, a voltage pulse is applied, using a Gaussian profile,

$$V_g(t) = V_{g,0}\left[1 \pm \exp\left(-\frac{t^2}{\sigma^2}\right)\right],$$

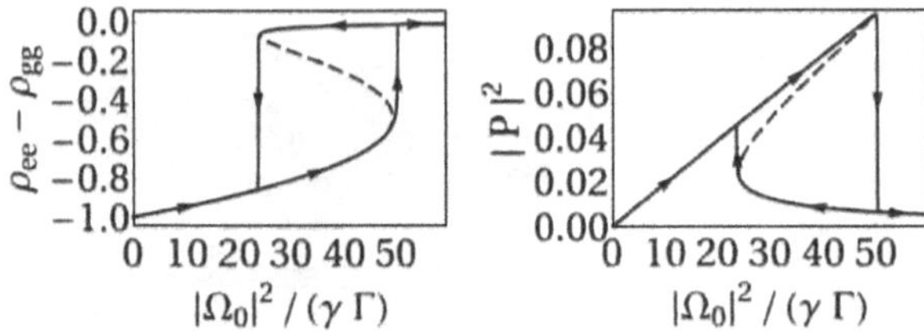

Fig. 7.7 Hysteresis loops (*solid lines*) for the population difference Z and the squared dipole amplitude $|P|^2$, when the external incoming field is swept adiabatically back and forth across the bistable region (arrows show the direction of sweeping), with a constant gate voltage $V_g = 0$. The stationary solutions are plotted using *dashed black lines* (the same line style as in Fig. 7.6)

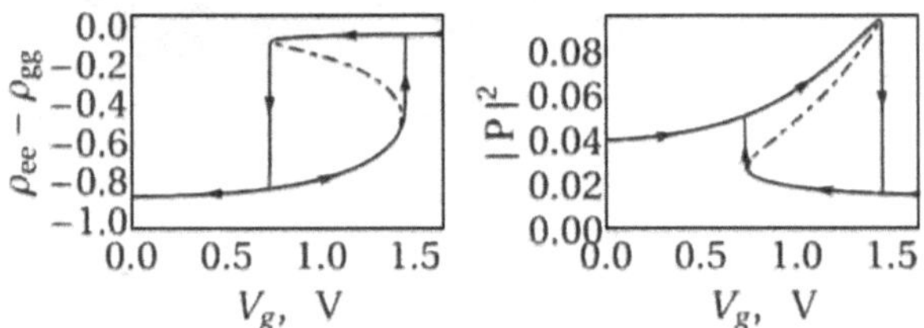

Fig. 7.8 Hysteresis loops (*solid lines*) for the population difference Z and the squared dipole amplitude $|P|^2$, when the gate voltage is swept adiabatically back and forth across the bistable region (arrows show the direction of sweeping), with a constant incoming field $|\Omega_0|^2/(\gamma\Gamma) = 17$. The stationary solutions are plotted using *dot-dashed black lines* (the same line style as in Fig. 7.6)

where σ defines the pulse duration. The plus sign applies for flips from the lower to the upper branch, whereas the minus is set for the opposite branch flip. The evolution of the system, for $2\sigma = 1, 2, 3$ ns, is plotted in Fig. 7.9. For $2\sigma = 1$ ns, the flipping does not occur in any direction, whereas for $2\sigma = 3$ ns it always happens. However, there is an intermediate regime, exemplified by $2\sigma = 2$ ns, where the system can only flip from the lower to the upper branch.

The dynamics of the system out of equilibrium is governed by the position of the stationary solution. The stable branches act as attractors, and the unstable as repeller. Therefore, after switching off the pulse, the system will prefer to move to the stable branch that can be reached without crossing the instability. This is apparent in the plots of the population: before reaching the initial voltage, the evolution bends in the opposite direction to the unstable branch.

However, more features become evident after a careful inspection of Fig. 7.9. The dynamics of the flips with positive voltage pulse (red lines) is faster than the one with negative pulses (blue). This question has been thoroughly addressed for an SQD+MNP system with optical hysteresis in [26]. The authors proved that the switching time from the lower to the upper branch is proportional to γ, and that the opposite switching time diverges if the excitation does not move far from the equilibrium position. For the system studied here, the critical point for this switching lies at $V_g \simeq 0.6$ V. Although the negative pulse reaches $V_g = 0$ V, which might seem

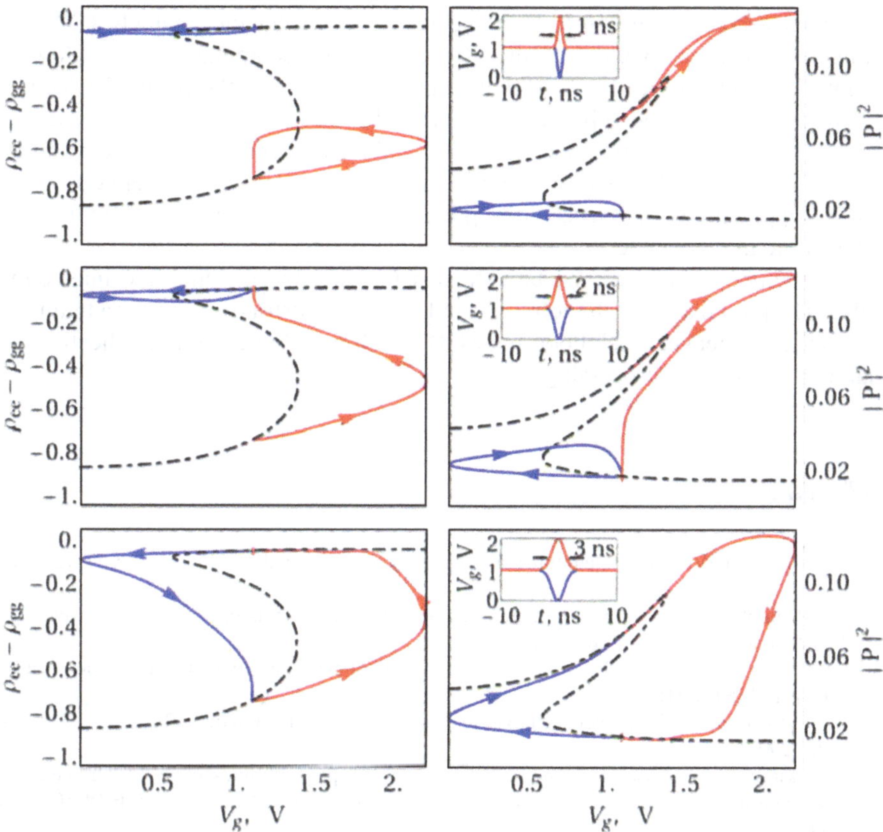

Fig. 7.9 Time evolution of the system under gate voltage pulses of different durations. *Dot-dashed lines* represent the stationary solutions. The gate-voltage pulse has a Gaussian profile $V_g(t) = 1.1[1 \pm \exp(-t^2/\sigma^2)]$ V, with $\sigma = 0.5$ ns (*upper panels*), $\sigma = 1.0$ ns (*central*) and $\sigma = 1.5$ ns (*lower*). *Blue lines* correspond to an initial condition in the upper stable branch which is subjected to a negative voltage pulse, whereas red ones start in the lower stable branch and are subjected to a positive pulse. Insets in the *right panels* show the pulses shape, along with the characteristic pulse duration 2σ

a large shift, the evolution of the feedback parameter G with the applied voltage is smooth (right panel of Fig. 7.2), and the renormalization of $\mathbf{E}_{0,\text{in}}$ is negligible, and therefore the net effect on the SQD is small.

7.5 Summary

A SQD embedded in a dielectric host grown on top of a substrate material was studied. The redistribution of charges at the interface, due to the dipole moment of the SQD results in a feedback mechanism for the SQD. It was shown that this self-action results in bistability and optical hysteresis.

The substrate was chosen to be a degenerate semiconductor, namely ITO. The application of a gate voltage to the ITO creates an accumulation layer close to the interface, which locally changes the optical response of the material [17], and therefore modulates the self-action of the SQD. Using that gate voltage, it was proved that for a fixed external illumination, the hysteresis of the system can be electrically controlled, providing a novel mechanism to control quantum dots, with possible applications such as electro-optical switches, modulators or memory cells at nanoscale in the visible.

To conclude, this work can be extended to systems with multiple interacting SQDs. The modulation of the refractive index in the substrate allows for a tuning of the interaction between neighbouring SQDs, with possible exciting applications in quantum information processing.

References

1. D. Loss, D.P. DiVincenzo, Quantum computation with quantum dots. Phys. Rev. A **57**, 120–126 (1998)
2. J.R. Petta et al., Coherent manipulation of coupled electron spins in semiconductor quantum dots. Science **309**, 2180–2184 (2005)
3. J. Elzerman et al., Single-shot read-out of an individual electron spin in a quantum dot. Nature **430**, 431–435 (2004)
4. W.G. van der Wiel et al., Electron transport through double quantum dots. Rev. Mod. Phys. **75**, 1–22 (2002)
5. A.I. Ekimov et al., Absorption and intensity-dependent photoluminescence measurements on cdse quantum dots: assignment of the first electronic transitions. J. Opt. Soc. Am. B **10**, 100–107 (1993)
6. D.J. Norris, M.G. Bawendi, Measurement and assignment of the size dependent optical spectrum in cdse quantum dots. Phys. Rev. B **53**, 16338–16346 (1996)
7. P. Hawrylak, M. Korkusiński, Electronic properties of self-assembled quantum dots. Single Quant. Dots **90**, 25–92 (2003)
8. C. Simon et al., Quantum memories. Eur. Phys. J. D: At. Mol. Opt. Plasma Phys. **58**, 1–22 (2010)
9. M. Kroutvar et al., Optically programmable electron spin memory using semiconductor quantum dots. Nature **432**, 81–84 (2004)
10. R.J. Young et al., Single electron-spin memory with a semiconductor quantum dot. New J. Phys. **9**, 365 (2007)
11. W. Zhang et al., Semiconductor-metal nanoparticle molecules: hybrid excitons and the nonlinear fano effect. Phys. Rev. Lett. **97**, 146804 (2006)
12. R.D. Artuso, G.W. Bryant, Optical response of strongly coupled quantum dot-metal nanoparticle systems: double peaked fano structure and bistability. Nano Lett. **8**, 2106–2111 (2008)
13. S.M. Sadeghi, Plasmonic metaresonances: molecular resonances in quantum dot–metallic nanoparticle conjugates. Phys. Rev. B **79**, 233309 (2009)
14. S.M. Sadeghi, Tunable nanoswitches based on nanoparticle meta-molecules. Nanotechnology **21**, 355501 (2010)
15. S.M. Sadeghi, Gain without inversion in hybrid quantum dot-metallic nanoparticle systems. Nanotechnology **21**, 455401 (2010)
16. A.V. Malyshev, V.A. Malyshev, Optical bistability and hysteresis of a hybrid metal-semiconductor nanodimer. Phys. Rev. B **84**, 035314 (2011)

17. E. Feigenbaum et al., Unity-order index change in transparent conducting oxides at visiblefrequencies. Nano Lett. **10**, 2111–2116 (2010)
18. A.L. Efros, Luminescence polarization of cdse microcrystals. Phys. Rev. B **46**, 7448–7458 (1992)
19. S. Empedocles et al., Three-dimensional orientation measurements of symmetric single chromophores using polarization microscopy. Nature **399**, 126–130 (1999)
20. A.I. Chizhik et al., Excitation isotropy of single cdse/zns nanocrystals. Nano Lett. **11**, 1131–1135 (2011)
21. L. Allen, J. Eberly, *Optical Resonance and Two-Level Atoms*. (Dover, New York, 1975)
22. C.F. Bohren, D.R. Huffman, *Absorption and Scattering of Light by Small Particles*. (Wiley, New York, 1983)
23. M. Paulus et al., Accurate and effcient computation of the green's tensor for stratified media. Phys. Rev. E **62**, 5797–5807 (2000)
24. M. Kildemo et al., Approximation of re ection coeficients for rapid real-time calculation of inhomogeneous films. J. Opt. Soc. Am. A **14**, 931–939 (1997)
25. A.V. Malyshev, Condition for resonant optical bistability. Phys. Rev. A **86**, 065804 (2012)
26. B.S. Nugroho et al., Bistable optical response of nanoparticle heterodimer: mechanism, phase diagram, and switching time. ArXiv e-prints **1209**, 3255 (2012)
27. R. Friedberg et al., Mirrorless optical bistability condition. Phys. Rev. A **39**, 3444–3446 (1989)
28. A.J. Ramsay et al., Damping of exciton rabi rotations by acoustic phonons in optically excited InGaAs/GaAs quantum dots. Phys. Rev. Lett. **104**, 017402 (2010)

Chapter 8
Conclusions and Prospects

In this chapter the conclusions of this Thesis are put together and summarized. A critical analysis will be made, exploring the possible limitations of our proposals. Finally, some currently ongoing research, as well as prospects, will be commented.

8.1 Conclusions Regarding Electronic Nanodevices Based on Graphene

In Chap. 3, a new quantum interference device based on a graphene nanoring with 60° turns was proposed and studied. Transport properties of the device were found to be very sensitive to the type of edges (zigzag or armchair). The ring comprised of nanoribbons with armchair edges and parabolic dispersion relation with a gap proved to be the most advantageous for electronic transport because, in that case, the transmission pattern presented wide bands of high electronic transmittance. It was shown that the current flow through the device can be controlled by the side-gate voltage. Such a voltage changes the relative phase of the electron wave function in the two arms of the ring resulting in constructive or destructive interferences at the drain. Consequently, it was proved that the current flow can be modulated efficiently without applying a magnetic field, so the device operates as a quantum interference effect transistor. Its performance was shown to be robust under moderate edge disorder. It should be noted that the most prominent feature of this setup, 60° turns, has been experimentally observed [1].

The former device was extended in Chap. 4 to be used as a novel spin filter, by placing a graphene ring with the same geometry above a ferromagnetic strip. It was shown that, due to the exchange splitting induced by the magnetic ions of the ferromagnetic layer, the transmission coefficient is different for spin up and spin down electrons, giving rise to the polarization of the conductance and the electric current. The ring geometry strongly enhances the current polarization, compared to a simple aGNR with a ferromagnetic layer on top of it, and it allows the current

J. Munárriz Arrieta, *Modelling of Plasmonic and Graphene Nanodevices*,
Springer Theses, DOI: 10.1007/978-3-319-07088-9_8,

and its polarization to be controlled by a side-gate voltage. A detailed study was made regarding the effect of edge disorder and other fabrication imperfections, such as the asymmetry of the ring. The predicted effects were shown to be robust under moderate perturbations. Under some circumstances, asymmetries in the ring enhance the behaviour of the polarization, particularly in the case where one of the arms is made wider than the other. Finally, a simple model was introduced to explain the observed Fano resonances in the transmissions, as well as their relative width.

The possibility of getting a spin-dependent response using ferromagnetic strips led to the idea of designing a spintronic device based on graphene nanoribbons. This would strongly simplify the fabrication process. However, an aGNR with a ferromagnetic strip does not provide a strong spin-dependent response. That is why an aGNR and a regular array of ferromagnetic strips grown on top of it was proposed and studied in Chap. 5. This regular array creates a spin-dependent superlattice. It was shown that the electric current through the device can be highly polarized. Moreover, the two polarized components of the current manifest non-monotonic dependencies on the source-drain voltage. In particular, both spin-dependent current-voltage characteristics present regions with negative differential resistance for the source-drain voltage in the range of a few millivolts. The device operates therefore as a low-voltage regular Esaki diode for the spin-polarized currents. The usage of a superlattice induced by ferromagnets rather than by usual electrostatic gates is very attractive from the point of view of the circuit integration and device density. Unlike the long-range electrostatic gate potentials which can interfere with each other, setting a practical lower limit for the inter-device distance, the exchange interaction is very short-ranged. Its characteristic length scale is on the order of one monolayer, which allows for very close packing of circuits and, consequently, considerably higher device densities. Finally, it was shown that the current polarization is also a non-monotonic function of the source-drain voltage in our proposed device, which makes it an Esaki spin diode. This is a key feature, as in a true spintronic device the degree of freedom that carries information is the polarization of the current rather than its magnitude.

All of the above designs must be operating in the single mode regime in order to use the interference effects. When the second mode comes into play the interference bands smear out and the current control is expected to be less efficient. In this regard, the dispersion relations of the nanoribbons forming the devices provide an important starting point because they define the appropriate energy window where one single mode is contributing to the transport. As the width of the nanoribbon increases the window shrinks while its lower edge approaches the Dirac point. As an example, in Chaps. 3 and 4, which used ribbons of width $w \simeq 15$ nm, such a window was on the order of 40 meV, whereas in Chap. 5 ($w \simeq 10$ nm), it widened to about 60 meV. On the other hand, electronic transport through wider nanoribbons is less affected by the edge disorder. These considerations should be taken into account when designing and fabricating the real world devices.

In this Thesis, charge carriers are treated as non-interacting particles. The role of interactions in graphene is a very active research field, with special focus on electron-electron and electron-phonon interactions. However, the inclusion of these effects lies beyond the scope of this Thesis. Moreover, the non-interacting picture has undergone a number of tests to compare with experimental measurements, with satisfactory results.

There is also a concern regarding the feasibility of the deposition of ferromagnets above (or below) graphene nanoribbons. It is difficult to state to which extent the electric properties of graphene are affected by the presence of the ferromagnet. However, a recent experimental work [2] suggests that this deposition does not strongly alter the structure of graphene, which shows a minor decrease in the mobility.

Finally, it should be pointed out that the presence of disorder in any graphene-based device is unavoidable. Electric measurements performed in graphene nanoribbons show I–V characteristics with strong resonances close to the neutrality point, as well as conductance gaps larger than the energy bands calculated theoretically [3]. The latter is a feature that can be mapped to a transport mechanism dominated by variable range hopping between localized state, and cannot be accounted for by simply removing atoms from the sample edges. However, this is a field in constant progress, and the fabrication of samples with increasing edge regularity has led to devices where quantum transport dominates [4].

8.2 Conclusions Regarding Electro-Optical Nanodevices

The radiation patterns of a plasmonic antenna were studied in Chap. 6. The antenna comprised a regular linear chain of identical metal nanospheres in close proximity to an interface between two media with high dielectric contrast. When such a system is illuminated from the higher refractive index side of the interface it can be excited by evanescent waves. In this case the excitation does not mask the useful antenna signal, which is advantageous for measurements and applications. It was proved that the radiation pattern and its directivity can be controlled by changing the incidence angles and/or polarization of the excitation. In particular, for some excitation geometries, the antenna pattern is characterized by a very narrow main lobe whose direction can be changed abruptly by a relatively small change of incidence angles.

Initially, the antenna radiation patterns were calculated using the traditional, much more numerically elaborated and accurate, Sommerfeld integrals approach. Then a much simpler one, based on the image dipole and the stationary phase approximations, was proposed. It was shown that, despite the complexity of the system, the analytical expression for the radiation patterns gives an excellent description of the main features of the antenna response. These simple formulas can become a useful tool for solving the inverse problem: engineering a system that has a desired radiation pattern.

Deviations from regularity in the size and shape of the particles and in the inter-sphere distance are to be expected in the experimental realization of this proposal. It is quite clear that disorder would lead to some smearing of the sharp features. However, plasmon resonances of nanoparticles are broad and a relatively small change of their sizes would not affect the polarizability to a large extent. The same can be said about the interaction: being a long range one it is not very sensitive to small deviations of particle positions.

In Chap. 7, an SQD embedded in a dielectric host grown on top of a substrate material was studied. The redistribution of bonded charges and currents at the interface, due to the dipole moment of the quantum dot results in a feedback mechanism for the SQD. It was shown that this self-action results in bistability and optical hysteresis. The substrate was chosen to be Indium Tin Oxide, a degenerate semiconductor. The application of a gate voltage to the ITO creates an accumulation layer close to the interface, which locally changes the optical response of the material [5], and therefore modulates the self-action of the SQD. Using that gate voltage, it was proved that for a fixed external illumination, the hysteresis of the system can be electrically controlled, providing a novel interaction mechanism with quantum dots, with possible applications such as electro-optical switches, modulators or memory cells at nanoscale in the visible.

In order to keep the number of parameters as small as possible, a number of assumptions were made in this work. First, the SQD was assumed to have a spherical symmetry. However, real CdSe-based SQDs feature a small ellipticity. In addition to it, its wurtzite structure produces an emission transition dipole moment which is degenerate only in two dimensions [6] (the excitation transition dipole moment remains 3-D isotropic). This feature could easily be implemented in the calculations but, for clarity, the simpler model was chosen.

The presence of an accumulation layer in the substrate is due to an applied back voltage. This requires adding a couple of leads to the system, which could be incompatible with the excitation via plane waves. However, it would be straightforward to implement a different type of illumination, e.g., via the guided modes of the structure [5].

8.3 Prospective Research

During this Thesis, a number of projects related to the ones presented here have been envisioned. A non-exhaustive list of them is:

- As an extension of Chaps. 3 and 4, it would be interesting to study the effect of a constant magnetic field in the ring devices. A system with both electric and magnetic fields could show some very interesting fundamental properties, such as electrically suppressed Aharonov-Bohm oscillations.
- The geometry in Chap. 5 was chosen to be symmetrical with respect to the axis of the nanoribbon. This simplified the usage of the Dirac formalism, as well as the analysis of the superlattice created. However, devices comprising asymmetrically placed ferromagnets could induce very exotic outputs. As an example, if the output carriers were angle-resolved, an angle-dependent spin polarization could be observed. Spin puddles, i.e., regions with higher probability of finding carriers with one spin sign, could also be found.

- Due to the limitations of the point dipole approximation, MNPs in Chap. 6 were restricted to be at a distance $d \geq 3r$ for neighbouring particles, and the centres were at a distance $z \geq 3r$ from the interface. If these limitations were avoided by using a more advanced numerical technique, new fascinating phenomena could be added to the system. As an example, particles placed closer to the interface are known to produce strong shifts of the resonances. Furthermore, if the bounds to the centre-to-centre distance were removed, multipolar contributions are to be expected, which would enrich the radiation patterns with new features. The replacement of the spheres by ellipsoids would increase the band width of the device.
- In the literature, it is widely accepted that the point dipole approximation holds for centre-to-centre distance $d \leq 3r$ [7, 8]. However such a criterion does not exist for the centre-to-interface distance. By studying the dipole moment with the Boundary Element Method, evidence was found that, for the parameters used in Chap. 6, the criterion is $z \gtrsim 3r$. However, a general criterion, regardless of the materials chosen, would be a fundamental breakthrough in this field.
- The work presented in Chap. 7 can be easily extended to systems with multiple interacting quantum dots. The modulation of the refractive index in the substrate allows for a tuning of the interaction between neighbouring SQDs, with possible exciting applications in quantum information processing.

References

1. X. Li et al., Chemically derived, ultrasmooth graphene nanoribbon semiconductors. Science **319**, 1229–1232 (2008)
2. A.G. Swartz et al., Integration of the ferromagnetic insulator EuO onto graphene. ACS Nano **6**, 10063–10069 (2012)
3. M.Y. Han et al., Electron transport in disordered graphene nanoribbons. Phys. Rev. Lett. **104**, 056801 (2010)
4. X. Wang et al., Graphene nanoribbons with smooth edges behave as quantum wires. Nat. Nanotechnol. **6**, 563–567 (2011)
5. E. Feigenbaum et al., Unity-order index change in transparent conducting oxides at visible frequencies. Nano Lett. **10**, 2111–2116 (2010)
6. A.L. Efros, Luminescence polarization of CdSe microcrystals. Phys. Rev. B **46**, 7448–7458 (1992)
7. S.Y. Park, D. Stroud, Surface-plasmon dispersion relations in chains of metallic nanoparticles: an exact quasistatic calculation. Phys. Rev. B **69**, 125418 (2004)
8. J.-Y. Yan et al., Optical properties of coupled metal-semiconductor and metalmolecule nanocrystal complexes: role of multipole effects. Phys. Rev. B **77**, 165301 (2008)

Appendix A
TMM and QTBM Methods

Throughout part I, the following problem has been put forward a number of times:

> Given a discrete system governed by a TB Hamiltonian, and connected to $N = 2$ semi-infinite leads, calculate the wave function $|\psi\rangle$ of the whole system, if the set of eigenmodes entering in the system through the leads is known.

Only the easiest case needed to be considered, i.e., a mode $|v_{S,n}(E)\rangle$ with a given wave number $k_{S,n}$ entered the system through the source lead, and no modes entered through the drain lead. Then, the wave function in the whole system that matched the boundary conditions was calculated. Using this wave function, it was possible to calculate the outgoing modes, both in the left lead—*reflection*—and in the right *transmission*.

This type of systems is a subset of the one sketched in Fig. A.1, comprising *N* leads, labelled $\mathcal{L}_i$, $i = 1, \ldots, N$. Each lead $\mathcal{L}_i$ is a quasi-one dimensional system with spatial period Δ_i, made up of identical cells, named $\mathcal{C}_{i,j}$, $j = 0, 1, \ldots$. The number of atoms in each cell is M_i. In the method described below, the atoms belonging to the cell $\mathcal{C}_{i,j}$ can only be connected to the atoms in the previous, in the same or in the next cell, i.e., $\mathcal{C}_{i,j-1}$, $\mathcal{C}_{i,j}$, $\mathcal{C}_{i,j+1}$. In order to fulfil this condition, the cell can span multiple primitive cells of the quasi-one dimensional lead. It is the goal of the present appendix to develop a method to calculate the scattering states of the system, extending the method considered in [1] to systems with non-rectangular lattices. The appendix is structured as follows. First, the QTBM is formally presented in Sect. A.1. This method requires the calculation of the eigenmodes corresponding to the leads, a topic covered in Sects. A.2 and A.3, and illustrated with simple examples.

A.1 Quantum Transmission Boundary Method

First, the ith lead is considered as an isolated system. At any given energy E, there are $2M_i$ eigenmodes, comprising:

J. Munárriz Arrieta, *Modelling of Plasmonic and Graphene Nanodevices*, Springer Theses, DOI: 10.1007/978-3-319-07088-9,

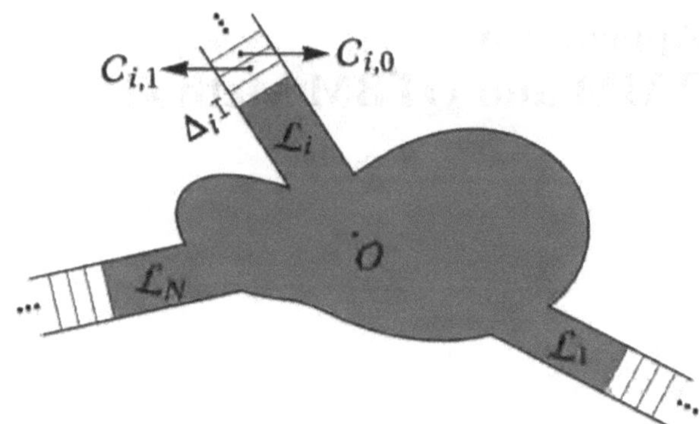

Fig. A.1 Scheme of the system where the QTBM is applied. It comprises N rectangular, perfect leads of infinite length, labelled $\mathcal{L}_i$. Each lead i can be split in sets of atoms of identical size M_i, labelled (from the outermost to the innermost), $\mathcal{C}_{i,j}$, $j = 0, 1, \ldots$, each with the same width Δ_i

- M_i ingoing eigenmodes $\left|\mathbf{v}_{i,n}^{\text{in}}\right\rangle$, with wavenumber $k_{i,n}^{\text{in}}$,
- M_i outgoing eigenmodes $\left|\mathbf{v}_{i,n}^{\text{out}}\right\rangle$, with wavenumber $k_{i,n}^{\text{out}}$,

where $\left|\mathbf{v}_{i,n}^{\text{in(out)}}\right\rangle$ is defined as the wave vector at $\mathcal{C}_{i,0}$. If the wave vector $k_{i,n}^{\text{in/out}}$ is real, then the eigenmode is propagating, otherwise being evanescent. For propagating eigenmodes, the test to see if it is going in or out of the system is the sign of its associated current (see Sect. 2.3.2), whereas for evanescent eigenmodes, they are considered ingoing if their amplitude decrease as they approach the system. It should be noted that this classification, though physically meaningful, is arbitrary: the formalism could also be applied to $2M_i$ *generic* eigenmodes, without further classifications.

The wave functions in two consecutive cells of the ith lead can be written in terms of the eigenmodes:

$$\begin{aligned}
\left|\mathcal{C}_{i,0}\right\rangle &= \sum_{n=1}^{N_i} a_{i,n}\left|\mathbf{v}_{i,n}^{\text{in}}\right\rangle + \sum_{n=1}^{N_i} b_{i,n}\left|\mathbf{v}_{i,n}^{\text{out}}\right\rangle \equiv \hat{\mathbf{V}}_i^{\text{in}}\cdot\mathbf{a}_i + \hat{\mathbf{V}}_i^{\text{out}}\cdot\mathbf{b}_i \\
\left|\mathcal{C}_{i,1}\right\rangle &= \sum_{n=1}^{N_i} a_{i,n}\left|\mathbf{v}_{i,n}^{\text{in}}\right\rangle e^{ik_{i,n}^{\text{in}}\Delta_i} + \sum_{n=1}^{N_i} b_{i,n}\left|\mathbf{v}_{i,n}^{\text{out}}\right\rangle e^{ik_{i,n}^{\text{out}}\Delta_i} \equiv \hat{\mathbf{W}}_i^{\text{in}}\cdot\mathbf{a}_i + \hat{\mathbf{W}}_i^{\text{out}}\cdot\mathbf{b}_i\,,
\end{aligned} \tag{A.1}$$

where the following matrix notation, with vectors aligned in columns, has been used:

$$\left(\hat{\mathbf{V}}_i^{\text{in(out)}}\right)_{m,n} \equiv \left(\mathbf{v}_{i,n}^{\text{in(out)}}\right)_m \quad \left(\hat{\mathbf{W}}_i^{\text{in(out)}}\right)_{m,n} \equiv e^{ik_{i,n}^{\text{in(out)}}\Delta_i}\left(\mathbf{v}_{i,n}^{\text{in(out)}}\right)_m\,. \tag{A.2}$$

In the problem considered, the amplitudes of the incoming modes $a_{i,n}$ are known, and the wave functions need to be calculated. Therefore, it is useful to use the above equations to remove the unknown outgoing amplitudes $b_{i,n}$:

$$\left|\mathcal{C}_{i,1}\right\rangle - \hat{\mathbf{W}}_i^{\mathrm{out}} \left(\hat{\mathbf{V}}_i^{\mathrm{out}}\right)^{-1} \left|\mathcal{C}_{i,0}\right\rangle = \left[\hat{\mathbf{W}}_i^{\mathrm{in}} - \hat{\mathbf{W}}_i^{\mathrm{out}} \left(\hat{\mathbf{V}}_i^{\mathrm{out}}\right)^{-1} \hat{\mathbf{V}}_i^{\mathrm{in}}\right] \cdot \mathbf{a}_i \ , \tag{A.3}$$

which form a set of N_i equations with $2N_i$ unknowns (the wave function in the atoms belonging to both cells).

Setting M to be the total number of atoms in the system, i.e., the total number of unknowns, M independent equations are needed:

- The time-independent Schrödinger equation

$$\mathcal{H} \left|\psi\right\rangle = E \left|\psi\right\rangle \tag{A.4}$$

  provides $M - \sum_{i=1}^{N} M_i$ equations. They can be obtained by projecting the wave function in any given atomic orbital $\left|\phi_j\right\rangle$, i.e., left multiplying $\left\langle\phi_j\right|$ at both sides of Eq. (A.4). This cannot be done for the atoms belonging to the first cell of each lead $\mathcal{C}_{i,0}$, as the Hamiltonian $\mathcal{H}$ is ill-defined: the atoms in these cells are connected to atoms outside the section of the system chosen as basis set.
- Equation (A.3) provides $\sum_{i=1}^{N} M_i$ equations.

It is possible to write this set of equations in a compact way. Starting with

$$(\mathcal{H} - E) \left|\psi\right\rangle \equiv \hat{\mathbf{M}}_0 \left|\psi\right\rangle = \mathbf{0} \ , \tag{A.5}$$

the rows of $\hat{\mathbf{M}}_0$ which correspond to atoms belonging to the cells $\mathcal{C}_{i,0}$ are replaced by the left side of the corresponding equation in Eq. (A.3). The element of the vector $\mathbf{0}$ in the same row is replaced by the independent term of the same equation, i.e., its right-hand side. Therefore, the scattering state $\left|\psi\right\rangle$ is obtained by solving a linear system,

$$\hat{\mathbf{M}} \left|\psi\right\rangle = \mathbf{b} \ , \tag{A.6}$$

with both $\hat{\mathbf{M}}$ and $\mathbf{b}$ being sparse arrays. Numerous numerical algorithms are available for solving this type of systems.

A.2 Eigenmodes in the Leads: Transfer Matrix Method

In the literature, the eigenmodes in the system for a given energy E are usually obtained by means of the Transfer Matrix Method [2]. Let $\hat{H}_{l,l-1}$, $\hat{H}_{l,l}$, $\hat{H}_{l,l+1}$ be the blocks of the full Hamiltonian for the lead, $\hat{H}$, which connect the atoms in the cell l with the atoms in the previous, in the same and in the next cells, respectively. If the Hamiltonian is projected on the basis formed by the atoms belonging to the lth cell, the following matrix equation is obtained:

$$\hat{H}_{l,l-1}\psi_{l-1} + \hat{H}_{l,l}\psi_l + \hat{H}_{l,l+1}\psi_{l+1} = E\psi_l \ , \tag{A.7}$$

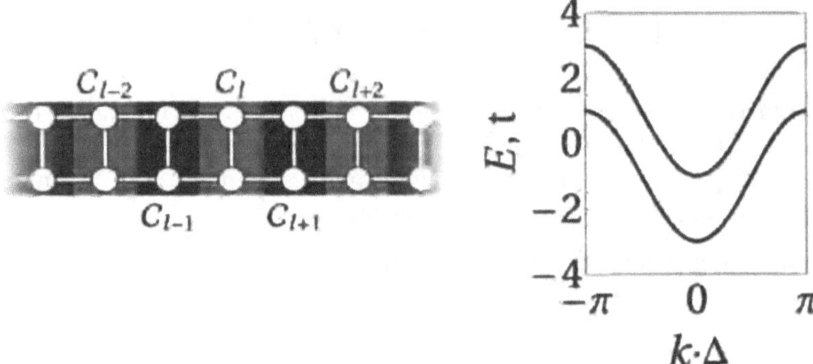

Fig. A.2 *Left panel* shows the scheme of an infinite ribbon with *square lattice*. For simplicity, the number of rows is set to 2. The lth cell and their neighbouring cells are highlighted using alternate *grey tones*. *Right panel* sketches the dispersion relation of the system

where ψ_l is the set of coefficients of the wave function in that basis. This equation can be rewritten as the following block matrix:

$$\begin{pmatrix} \psi_{l+1} \\ \psi_l \end{pmatrix} = \begin{pmatrix} -\hat{H}_{l,l+1}^{-1}\left(\hat{H}_{l,l} - E\right) & -\hat{H}_{l,l+1}^{-1}\hat{H}_{l,l-1} \\ \hat{\mathbf{1}} & \hat{\mathbf{0}} \end{pmatrix} \begin{pmatrix} \psi_l \\ \psi_{l-1} \end{pmatrix} = \mathrm{e}^{ik\Delta} \begin{pmatrix} \psi_l \\ \psi_{l-1} \end{pmatrix}, \tag{A.8}$$

where Δ is the cell width and the latter equality has been obtained using Bloch's theorem. Therefore, the eigenmodes in Eq. (A.8) correspond to the propagating (or evanescent) eigenmodes in the structure.

However, this method requires a matrix inversion. If there are atoms in the cell that are not connected both to the following cell, then there will be rows of the matrix $H_{l,l+1}$ with all their elements equal 0, which results in a singular matrix. This is the case of graphene.

A.2.1 Application: Square Lattice

As a simple example of the TMM, an infinite ribbon with square lattice and two rows is considered (see Fig. A.2). Setting the hopping $t = -1$ and the onsite energy to 0, the part of the Hamiltonian which affects the lth cell reads

$$\begin{pmatrix} \ddots & \vdots & \vdots & \vdots & \cdot^{\cdot^{\cdot}} \\ \cdots & H_{l-1,l-1} & H_{l-1,l} & H_{l-1,l+1} & \cdots \\ \cdots & H_{l,l-1} & H_{l,l} & H_{l,l+1} & \cdots \\ \cdots & H_{l+1,l-1} & H_{l+1,l} & H_{l+1,l+1} & \cdots \\ \cdot^{\cdot^{\cdot}} & \vdots & \vdots & \vdots & \ddots \end{pmatrix} = - \left(\begin{array}{c|cc|cc|cc|c} \ddots & \vdots & \vdots & \vdots & \vdots & \vdots & \vdots & \cdot^{\cdot^{\cdot}} \\ \hline \cdots & 0 & 1 & 1 & 0 & 0 & 0 & \cdots \\ \cdots & 1 & 0 & 0 & 1 & 0 & 0 & \cdots \\ \hline \cdots & 1 & 0 & 0 & 1 & 1 & 0 & \cdots \\ \cdots & 0 & 1 & 1 & 0 & 0 & 1 & \cdots \\ \hline \cdots & 0 & 0 & 1 & 0 & 0 & 1 & \cdots \\ \cdots & 0 & 0 & 0 & 1 & 1 & 0 & \cdots \\ \hline \cdot^{\cdot^{\cdot}} & \vdots & \vdots & \vdots & \vdots & \vdots & \vdots & \ddots \end{array}\right) \tag{A.9}$$

The transfer matrix T can be constructed using the definition in Eq. (A.8):

$$T = \begin{pmatrix} -E & -1 & -1 & 0 \\ -1 & -E & 0 & -1 \\ 1 & 0 & 0 & 0 \\ 0 & 1 & 0 & 0 \end{pmatrix}. \tag{A.10}$$

The eigenvalues of this matrix are

$$\begin{aligned} \lambda_1 &= \frac{1}{2}\left(1 - E \pm \sqrt{-3 - 2E + E^2}\right) \\ \lambda_2 &= \frac{1}{2}\left(-1 - E \pm \sqrt{-3 + 2E + E^2}\right), \end{aligned} \tag{A.11}$$

and setting $\lambda_i = e^{ik\Delta}$, the following dispersion relations for the two bands are obtained:

$$E(k) = \pm 1 - 2\cos(k\Delta). \tag{A.12}$$

Note that the four allowed values in Eq. (A.11) lead to only two bands, due to the $k \leftrightarrow -k$ symmetry of the two possible sign choices of the square root. The band structure is plotted in the right panel of Fig. A.2. The same result can be obtained by using symmetric and antisymmetric *ansatz* wave functions, i.e., a wave function with the equal values $(1, 1)$ in the two atoms of the unit cell, or with opposite signs, $(1, -1)$.

A.3 Effective Transfer Matrix Method

A generalization of the TMM allows the eigenmodes of more general structures to be calculated (see [3], appendix A, for its application to graphene). It requires a careful analysis of the unit cell to divide it into two sub-cells, labelled L and R. The sites in both of them must only be connected to sites belonging to the same subcell or to its nearest neighbours. Furthermore, all the sites belong to L *must* be connected to both neighbouring R subcells.

Then, the time-independent Schrödinger equation reads

$$\begin{pmatrix} \ddots & \vdots & \vdots & \vdots & \vdots & \iddots \\ \cdots & h_L & s_{L,R} & 0 & 0 & \cdots \\ \cdots & s^*_{L,R} & h_R & d_{R,L} & 0 & \cdots \\ \cdots & 0 & d^*_{R,L} & h_L & s_{L,R} & \cdots \\ \cdots & 0 & 0 & s^*_{L,R} & h_R & \cdots \\ \iddots & \vdots & \vdots & \vdots & \vdots & \ddots \end{pmatrix} \begin{pmatrix} \vdots \\ \psi_{l,L} \\ \psi_{l,R} \\ \psi_{l+1,L} \\ \psi_{l+1,R} \\ \vdots \end{pmatrix} = E \begin{pmatrix} \vdots \\ \psi_{l,L} \\ \psi_{l,R} \\ \psi_{l+1,L} \\ \psi_{l+1,R} \\ \vdots \end{pmatrix}, \tag{A.13}$$

where the discrete translation symmetry of the infinite ribbon and the hermiticity of the Hamiltonian, have been used. It should be noted that h_L, h_R are square matrices, whereas $s_{L,R}$, $d_{R,L}$ are in general rectangular. Two different equations can be obtained from this eigenvalue equation:

$$\begin{aligned} s^*_{L,R}\psi_{n,L} + h_R\psi_{n,R} + d_{R,L}\psi_{n+1,L} &= E\ \psi_{n,R} \\ d^*_{R,L}\psi_{n,R} + h_L\psi_{n+1,L} + s_{L,R}\psi_{n+1,R} &= E\ \psi_{n+1,L}\ . \end{aligned} \tag{A.14}$$

It is possible to combine both equations to get one which only involves wave functions in the L sub-cells, but it is necessary to calculate left inverses of rectangular matrices. It can be shown that the left inverse of a matrix with dimensions $m \times n$ exists if and only if the rows (columns) are linearly independent, for $m > n (m < n)$ [4]. This requirement is fulfilled by setting the atoms to be connected as described before. Then:

$$\begin{aligned} & s_{L,R}\,(E - h_R)^{-1} d_{R,L}\psi_{n+2,L} \\ & \quad = \left(E - d^*_{R,L}\,(E - h_R)^{-1} d_{R,L} - s_{L,R}\,(E - h_R)^{-1} s^*_{L,R} - h_L\right)\psi_{n+1,L} \\ & \qquad - d^*_{R,L}\,(E - h_R)^{-1} s^*_{L,R}\psi_{n,L} \\ & \quad \Rightarrow \psi_{n+2,L} = \hat{T}_{1,1}\psi_{n+1,L} + \hat{T}_{1,2}\psi_{n,L}\ , \end{aligned} \tag{A.15}$$

which can be rewritten as a transfer matrix,

$$\begin{pmatrix} \psi_{n+2,L} \\ \psi_{n+1,L} \end{pmatrix} = \begin{pmatrix} \hat{T}_{1,1} & \hat{T}_{1,2} \\ \hat{\mathbf{1}} & \hat{\mathbf{0}} \end{pmatrix} \begin{pmatrix} \psi_{n+1,L} \\ \psi_{n,L} \end{pmatrix}. \tag{A.16}$$

A.3.1 Application: Hexagonal Lattice

- **aGNR with $N = 1$**

Using the previous approach, the dispersion relation of the simplest aGNR will be

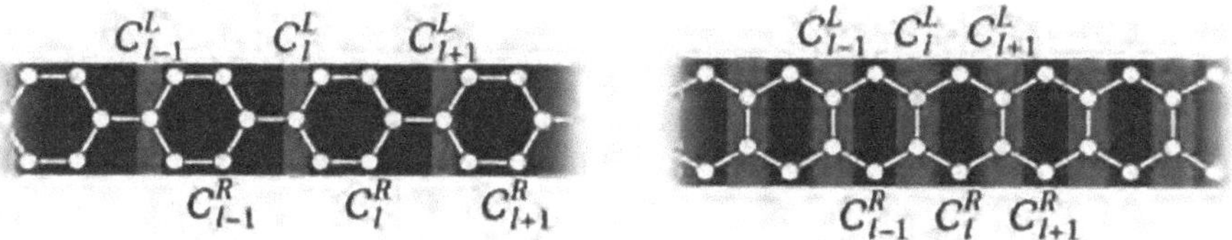

Fig. A.3 Scheme of an infinite aGNR (zGNR) is shown in the *left* (*right*) panel. For simplicity, the width of both samples has been set to one cell. A valid separation of the unit cells in two sub-cells is plotted using alternative *light grey*—*L* regions— and dark—*R*

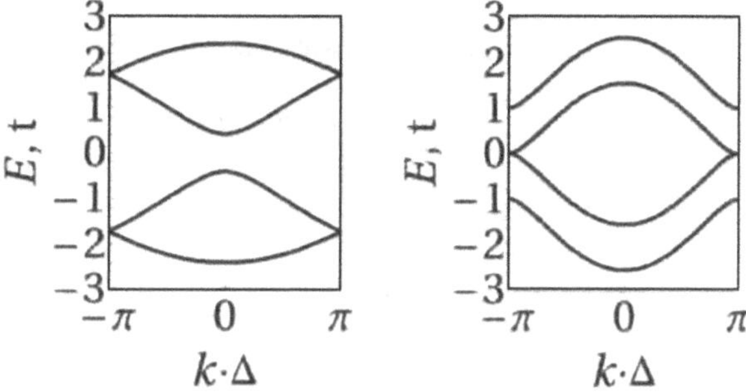

Fig. A.4 Dispersion relations of the nanoribbons shown in Fig. A.3, obtained using the eigenvalues of the matrices Eqs. (A.18) and (A.20)

studied. This ribbon is plotted in the left panel of Fig. A.3. The atoms are numbered from left to right, and if they belong to the same column, from the lower to the upper. The elements of the Hamiltonian defined by Eq. (A.13) read:

$$h_L = \begin{pmatrix} 0 \end{pmatrix} \qquad s_{L,R} = \begin{pmatrix} -1 & -1 & 0 & 0 & 0 \end{pmatrix}$$

$$h_R = \begin{pmatrix} 0 & 0 & -1 & 0 & 0 \\ 0 & 0 & 0 & -1 & 0 \\ -1 & 0 & 0 & 0 & -1 \\ 0 & -1 & 0 & 0 & -1 \\ 0 & 0 & -1 & -1 & 0 \end{pmatrix} \qquad d_{R,L} = \begin{pmatrix} 0 \\ 0 \\ 0 \\ 0 \\ -1 \end{pmatrix}.$$

Defining the effective transfer matrix T as in Eq. (A.16),

$$T = \begin{pmatrix} \frac{1}{2}\left(5 - 6E^2 + E^4\right) & -1 \\ 1 & 0 \end{pmatrix} \tag{A.18}$$

is obtained. Then, setting the eigenvalues to be $e^{ik\Delta}$, the dispersion relation is obtained, and plotted in the left panel of Fig. A.4. The symmetry $E \leftrightarrow -E$ is due to $\det(T)$ being a biquadratic polynomial of E. The number of bands being smaller than the number of atoms in the unit cell is due to the symmetric configuration, which produces degenerate bands.

- **zGNR with $N = 1$**

The same procedure can be applied to calculate the dispersion relation of a zGNR with $N = 1$. In this case, the elements of the Hamiltonian read:

$$h_L = \begin{pmatrix} 0 & -1 \\ -1 & 0 \end{pmatrix} \qquad s_{L,R} = \begin{pmatrix} -1 & 0 \\ 0 & -1 \end{pmatrix}$$
$$h_R = \begin{pmatrix} 0 & 0 \\ 0 & 0 \end{pmatrix} \qquad d_{R,L} = \begin{pmatrix} -1 & 0 \\ 0 & -1 \end{pmatrix} . \tag{A.19}$$

The resulting effective Transfer Matrix,

$$T = \begin{pmatrix} -2+E^2 & E & -1 & 0 \\ E & -2+E^2 & 0 & -1 \\ 1 & 0 & 0 & 0 \\ 0 & 1 & 0 & 0 \end{pmatrix}, \tag{A.20}$$

allows the dispersion relation to be obtained (right panel of Fig. A.4).

References

1. C.S. Lent, D.J. Kirkner, The quantum transmitting boundary method. Journal of Applied Physics **67**, 6353–6359 (1990)
2. D.Z.Y. Ting et al., Multiband treatment of quantum transport in interband tunnel devices. Physical Review B **45**, 3583–3592 (1992)
3. J. Schelter et al., Interplay of the Aharonov-Bohm effect and Klein tunneling in graphene. Physical Review B **81**, 195441 (2010)
4. C.M. Fernández, Condiciones para la existencia de matriz inversa de un matriz no cuadrada. Gaceta Matemática **26**, 125–126 (1974)

Appendix B
Green's Tensor in a Stratified Media

In the second part of this Thesis, corresponding to Chaps. 6 and 7, it has been necessary to calculate the radiation of a dipole on top of a stratified medium. This problem traces back more than a century, to the seminal work by Sommerfeld [1]. Originally, it was formulated to understand the effect of the Earth on the transmission of signals through *wireless telegraphy*. If the electromagnetic excitation is a plane wave, the problem can easily be solved by using the Fresnel coefficients, which connect the incident, reflected and transmitted waves with the appropriate boundary conditions. However, the case of dipoles is far more complex, with the algebra leading to final expressions which involve the integration of non-trivial functions in the complex plane.

Due to the central importance of this problem in different fields of science and engineering, a variety of rigorous treatments can be found in the literature [5–10]. In these references, the full problem is considered, i.e., the dipole and the measurement can be at any position within the stratified media. In this appendix, a simpler, more intuitive formalism will be provided, with both $\mathbf{r}_0$ and $\mathbf{r}$ lying in the upper semi-infinite medium (see Fig. B.1). Green's tensor (also called Green's dyadic), which relates the field at $\mathbf{r}$ with the source dipole $\mathbf{p}$ at $\mathbf{r}_0$ will be obtained.

The procedure followed is conceptually simple. A spectral decomposition of Green's dyadic in a homogeneous medium can be made, following the recipe of [3], decomposing the dipole radiation in an infinite set of propagating and evanescent waves. Every component of this decomposition can be forced to fulfil the boundary conditions at the interfaces, which are given by the Fresnel coefficients. The resulting expression will need to be numerically integrated.

J. Munárriz Arrieta, *Modelling of Plasmonic and Graphene Nanodevices*,
Springer Theses, DOI: 10.1007/978-3-319-07088-9,

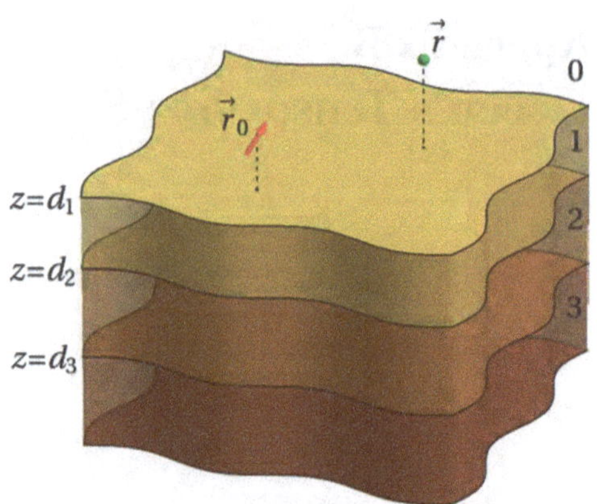

Fig. B.1 Schematics of the system studied: a dipole is placed in the *upper*, semi-infinite layer of an stratified system of $N+1$ layers, labelled from the *upper* to the *lower* $(0, \ldots, N)$, and N interfaces, labelled $1, \ldots, N$

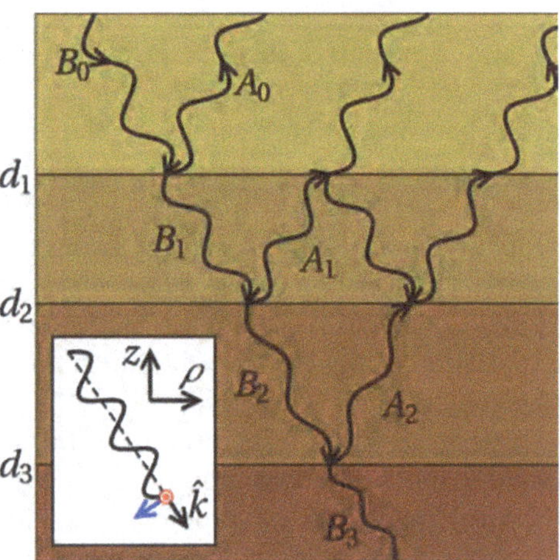

Fig. B.2 Schematics of the waves. For any layer l fulfilling $0 < l < N$, N being the number of interfaces, the up-going and down-going waves, with amplitudes labelled A_ℓ, B_ℓ, are formally constructed by the superposition of an infinite set of reflected and transmitted waves. The *lower left inplot* shows the polarization vectors $\hat{\mathbf{s}}$—*red*—and $\hat{\mathbf{p}}$—*blue*—for a down-propagating plane wave with unitary wave vector $\hat{\mathbf{k}}$

B.1 Reflection and Transmission of a Plane Wave Due to a Stratified Media

First, the behaviour of a plane wave arriving to the upper interface will be studied. The geometry considered here is given by a set of N interfaces, placed at $z = d_i$, $i = 1, \ldots, N$, separating $N+1$ different media. The properties of the media are given by their corresponding (scalar) permittivities and permeabilities, ϵ_ℓ, μ_ℓ. This constants determine the norm of the wave vector in each media, $k_\ell = (\omega/c)n_\ell$, with $n_\ell \equiv \sqrt{\epsilon_\ell \mu_\ell}$ being the refractive index.

Let a plane wave arrive to the the upper interface. The incidence of this excitation in the stratum produces a cascade of reflection and transmission events, as sketched in Fig. B.2. The energy and horizontal momentum conservation in these events lead to two important consequences, namely: (i) the components of the horizontal momentum $\mathbf{k}_\rho$ are conserved across the interface, and (ii) the amplitudes of the reflected and transmitted waves can be related via the Fresnel coefficients:

$$\mathcal{R}^s_{\ell,\ell+1} = \frac{\mu_{\ell+1}k_{\ell,z} - \mu_\ell k_{\ell+1,z}}{\mu_{\ell+1}k_{\ell,z} + \mu_\ell k_{\ell+1,z}} \qquad \mathcal{R}^p_{\ell,\ell+1} = \frac{\epsilon_{\ell+1}k_{\ell,z} - \epsilon_\ell k_{\ell+1,z}}{\epsilon_{\ell+1}k_{\ell,z} + \epsilon_\ell k_{\ell+1,z}} \tag{B.1}$$
$$\mathcal{T}^s_{\ell,\ell+1} = \frac{2\mu_{\ell+1}k_{\ell,z}}{\mu_{\ell+1}k_{\ell,z} + \mu_\ell k_{\ell+1,z}} \qquad \mathcal{T}^p_{\ell,\ell+1} = \frac{n_{\ell+1}}{n_\ell}\frac{2\epsilon_\ell k_{\ell,z}}{\epsilon_{\ell+1}k_{\ell,z} + \epsilon_\ell k_{\ell+1,z}},$$

with $k_{\ell,z} = (k_\ell^2 - k_\rho^2)^{1/2}$, and the superscripts s, p standing for waves with electric field polarized along $\hat{\mathbf{s}}$ (TE modes) and $\hat{\mathbf{p}}$ (TM modes; see the lower left inplot of Fig. B.2). If the wave vector is given by $\hat{\mathbf{k}} = (k_x, k_y, \pm k_{l,z})$, then

$$\hat{\mathbf{k}}(\mathbf{k}_\rho, k_{\ell,z}) = \frac{\mathbf{k}}{k} \quad \hat{\mathbf{s}}(\mathbf{k}_\rho, k_{\ell,z}) = \frac{\hat{\mathbf{k}} \times \hat{\mathbf{z}}}{|\hat{\mathbf{k}} \times \hat{\mathbf{z}}|} \quad \hat{\mathbf{p}}(\mathbf{k}_\rho, k_{\ell,z}) = \hat{\mathbf{k}} \times \hat{\mathbf{s}}\,. \tag{B.2}$$

Due to the conservation of horizontal momentum $\mathbf{k}_\rho$, the superposition of plane waves in each layer can be represented by a set of only two different waves for each polarization: the up-going one, given by $(\mathbf{k}_\rho, +k_{\ell,z})$, and the down-going, $(\mathbf{k}_\rho, -k_{\ell,z})$, which, for a point placed at $\mathbf{r} = (\rho, z)$, read:

$$\mathcal{A}_\ell(\mathbf{r}) = A_\ell e^{ik_{\ell,z}z} e^{i\mathbf{k}_\rho\cdot\rho}\hat{\mathbf{e}} \equiv \mathcal{A}_\ell(z)e^{i\mathbf{k}_\rho\cdot\rho}\hat{\mathbf{e}} \tag{B.3}$$
$$\mathcal{B}_\ell(\mathbf{r}) = B_\ell e^{-ik_{\ell,z}z} e^{i\mathbf{k}_\rho\cdot\rho}\hat{\mathbf{e}} \equiv \mathcal{B}_\ell(z)e^{i\mathbf{k}_\rho\cdot\rho}\hat{\mathbf{e}}\,,$$

where the indices s, p have been dropped, and $\hat{\mathbf{e}} = \hat{\mathbf{s}},\ \hat{\mathbf{p}}$ in each case. Note that the dephasing of the plane waves in each layer has been arbitrarily taken in such a way that it would be 0 at $\mathbf{r} = 0$ if all the space was filled with the medium in that layer. This induces an "artificial" dephasing between plane waves in neighbouring layers, which must be accounted for in the coefficients $A_\ell,\ B_\ell$.

The boundary conditions to be met at the ℓth interface are

$$\mathcal{B}_\ell(d_\ell) = \mathcal{R}_{\ell,\ell-1}\mathcal{A}_\ell(d_\ell) + \mathcal{T}_{\ell-1,\ell}\mathcal{B}_{\ell-1}(d_\ell) \tag{B.4}$$
$$\mathcal{A}_{\ell-1}(d_\ell) = \mathcal{R}_{\ell-1,\ell}\mathcal{B}_{\ell-1}(d_\ell) + \mathcal{T}_{\ell,\ell-1}\mathcal{A}_\ell(d_\ell)\,,$$

which can be written as a transfer matrix:

$$\begin{pmatrix} \mathcal{A}_\ell(d_\ell) \\ \mathcal{B}_\ell(d_\ell) \end{pmatrix} = \frac{1}{\mathcal{T}_{\ell,\ell-1}} \begin{pmatrix} 1 & \mathcal{R}_{\ell,\ell-1} \\ \mathcal{R}_{\ell,\ell-1} & 1 \end{pmatrix} \begin{pmatrix} \mathcal{A}_{\ell-1}(d_\ell) \\ \mathcal{B}_{\ell-1}(d_\ell) \end{pmatrix} \equiv \hat{\mathbf{s}}_\ell \begin{pmatrix} \mathcal{A}_{\ell-1}(d_\ell) \\ \mathcal{B}_{\ell-1}(d_\ell) \end{pmatrix}, \tag{B.5}$$

where the identities

$$\mathcal{R}_{\ell-1,\ell} = -\mathcal{R}_{\ell,\ell-1} \qquad \mathcal{T}_{\ell-1,\ell}\,\mathcal{T}_{\ell,\ell-1} - \mathcal{R}_{\ell-1,\ell}\,\mathcal{R}_{\ell,\ell-1} = 1\,,$$

have been used to simplify the expression. The propagation along the lth layer, from z_1 to z_2, can be directly obtained from Eq. (B.3), yielding:

$$\begin{aligned}
\begin{pmatrix} \mathcal{A}_\ell(z_2) \\ \mathcal{B}_\ell(z_2) \end{pmatrix} &= \begin{pmatrix} e^{ik_{\ell,z}(z_2-z_1)} & 0 \\ 0 & e^{-ik_{\ell,z}(z_2-z_1)} \end{pmatrix} \begin{pmatrix} \mathcal{A}_\ell(z_1) \\ \mathcal{B}_\ell(z_1) \end{pmatrix} \\
&= \begin{pmatrix} e^{ik_{\ell,z}z_2} & 0 \\ 0 & e^{-ik_{\ell,z}z_2} \end{pmatrix} \begin{pmatrix} e^{-ik_{\ell,z}z_1} & 0 \\ 0 & e^{ik_{\ell,z}z_1} \end{pmatrix} \begin{pmatrix} \mathcal{A}_\ell(z_1) \\ \mathcal{B}_\ell(z_1) \end{pmatrix} \\
&\equiv \hat{\mathbf{G}}_\ell^{-1}(z_2)\hat{\mathbf{G}}_\ell(z_1) \begin{pmatrix} \mathcal{A}_\ell(z_1) \\ \mathcal{B}_\ell(z_1) \end{pmatrix} ,
\end{aligned} \tag{B.6}$$

which finally allows the upper and lower layers of the stratum to be connected:

$$\begin{pmatrix} \mathcal{A}_N(z_N) \\ \mathcal{B}_N(z_N) \end{pmatrix} = \hat{\mathbf{G}}_N^{-1}(z_N) \left(\prod_{\ell=N}^{1} \hat{\mathbf{G}}_\ell(d_\ell)\hat{\mathbf{s}}_\ell\hat{\mathbf{G}}_{\ell-1}^{-1}(d_\ell) \right) \hat{\mathbf{G}}_0(z_0) \begin{pmatrix} \mathcal{A}_0(z_0) \\ \mathcal{B}_0(z_0) \end{pmatrix} \tag{B.7}$$

$$\equiv \hat{\mathbf{T}}(z_N, z_0) \begin{pmatrix} \mathcal{A}_0(z_0) \\ \mathcal{B}_0(z_0) \end{pmatrix} \tag{B.8}$$

From the elements of the matrix $\hat{\mathbf{T}}(z_N, z_0)$ of the latter expression, generalized transmitted and reflected coefficients can be derived by setting $\mathcal{A}_N(z_2) = 0$ (no plane wave coming from $z = -\infty$), $z_0 = d_1$ and $z_N = d_N$:

$$\mathcal{R}_{0,N} = -\frac{(\hat{\mathbf{T}})_{1,2}}{(\hat{\mathbf{T}})_{1,1}} \qquad \mathcal{T}_{0,N} = \frac{\det(\hat{\mathbf{T}})}{(\hat{\mathbf{T}})_{1,1}} \tag{B.9}$$

The validity of these expressions can be checked by comparing them with calculations using a formalism of multiple scattering. As an example, the reflection of a plane wave in a system with three media (two interfaces separated by a distance d), can be calculated by summing up the optical paths of rays that undergo a reflection in the first interface ($\propto \mathcal{R}_{0,1}$), initial transmission followed by a reflection in the second interface ($\propto \mathcal{T}_{0,1}e^{ik_1 d}\mathcal{R}_{1,2}e^{ik_1 d}\mathcal{T}_{1,0}$), etc. An analogous procedure can be used to calculate the transmission:

$$\begin{aligned}
\mathcal{R}_{0,2} &= \mathcal{R}_{0,1} + \mathcal{T}_{0,1}e^{ik_1 d}\mathcal{R}_{1,2}e^{ik_1 d}\mathcal{T}_{1,0} + \mathcal{T}_{0,1}e^{ik_1 d}\mathcal{R}_{1,2}e^{ik_1 d}\mathcal{R}_{1,0}e^{ik_1 d}\mathcal{R}_{1,2}e^{ik_1 d}\mathcal{T}_{1,0} + \cdots \\
&= \mathcal{R}_{0,1} + \mathcal{T}_{0,1}e^{ik_1 d}\mathcal{R}_{1,2} \left[\sum_{n=0}^{\infty} \left(e^{ik_1 d}\mathcal{R}_{1,0}e^{ik_1 d}\mathcal{R}_{1,2} \right)^n \right] e^{ik_1 d}\mathcal{T}_{1,0} \\
&= \mathcal{R}_{0,1} + \frac{e^{2ik_1 d}\mathcal{T}_{0,1}\mathcal{T}_{1,0}}{1 - e^{2ik_1 d}\mathcal{R}_{1,0}\mathcal{R}_{1,2}}
\end{aligned} \tag{B.10}$$

$$\begin{aligned}
\mathcal{T}_{0,2} &= \mathcal{T}_{0,1}e^{ik_1 d}\mathcal{T}_{1,2} + \mathcal{T}_{0,1}e^{ik_1 d}\mathcal{R}_{1,2}e^{ik_1 d}\mathcal{R}_{1,0}e^{ik_1 d}\mathcal{T}_{1,2} + \cdots \\
&= \frac{e^{ik_1 d}\mathcal{T}_{0,1}\mathcal{T}_{1,2}}{1 - e^{2ik_1 d}\mathcal{R}_{1,2}\mathcal{R}_{1,0}} ,
\end{aligned} \tag{B.11}$$

which are the results obtained using the transfer matrix method.

B.2 Electric Field Due to a Dipole on Top of a Substrate

The derivation of Green's tensor that relates the electric field at $\mathbf{r}$ with a source dipole at $\mathbf{r}_0$ has been presented in the literature in a number of ways [8, 9]. Here, a method based on the simultaneous calculation of the scalar and vector potentials has been chosen.

First, the Maxwell equations will be used to obtain independent expressions for the potentials in terms of the sources. In gaussian units and frequency domain, the equations read:

$$\begin{aligned} \nabla \cdot \mathbf{E} &= 4\pi \frac{\rho}{\epsilon} \qquad & \nabla \times \mathbf{H} + \frac{i\omega\epsilon}{c}\mathbf{E} &= \frac{4\pi}{c}\mathbf{j} \\ \nabla \cdot \mathbf{H} &= 0 \qquad & \nabla \times \mathbf{E} - \frac{i\omega\mu}{c}\mathbf{H} &= \mathbf{0}\,, \end{aligned} \tag{B.12}$$

where the dependence $e^{-i\omega t}$ has been assumed for all the quantities. For simplicity, $\mathbf{E}$ and $\mathbf{H}$ are derived in terms of potentials:

$$\mathbf{E} = \frac{i\omega}{c}\mathbf{A} - \nabla\phi \qquad \mathbf{H} = \frac{1}{\mu}\nabla \times \mathbf{A}\,. \tag{B.13}$$

By using the Lorenz gauge $\nabla \cdot \mathbf{A} = i\omega\epsilon\mu\phi/c$, it is possible to write two separate differential equations for the potentials [7]:

$$\left(\nabla^2 + k^2\right)\phi = -4\pi\left(\frac{\rho}{\epsilon} + \frac{1}{4\pi}\mathbf{D}\cdot\nabla\frac{1}{\epsilon}\right) \tag{B.14}$$

$$\left(\nabla^2 + k^2\right)\mathbf{A} = -\frac{4\pi}{c}\left(\mu\mathbf{j} - \frac{1}{4\pi}\left[i\omega\phi\nabla\left(\epsilon\mu\right) + c\,\mathbf{H}\times\nabla\mu\right]\right)\,, \tag{B.15}$$

with $k^2 = (\omega^2/c^2)\epsilon\mu$ being the wave vector in the medium.

In a homogeneous and isotropic medium, both μ and ϵ are constant scalars, the second terms in the right side of Eqs. (B.14) and (B.15) vanish, and the potentials can be obtained using Green's function formalism,

$$\phi(\mathbf{r}) = \frac{1}{\epsilon}\int d\mathbf{r}' G_H(|\mathbf{r}-\mathbf{r}'|)\rho(\mathbf{r}') \qquad \mathbf{A}(\mathbf{r}) = \frac{1}{c}\int d\mathbf{r}' G_H(|\mathbf{r}-\mathbf{r}'|)\mathbf{j}(\mathbf{r}')\,, \tag{B.16}$$

where

$$G_H(r) = \frac{e^{ikr}}{r} \tag{B.17}$$

is Green's function in a homogeneous environment, given by the wave equation

$$\left[\nabla^2 + k^2\right]G_H(r) = -4\pi\delta(\mathbf{r})\,. \tag{B.18}$$

By using this definition for Green's function, it is necessary to have k values yielding positive imaginary parts, so that Green's function vanishes as $r \to \infty$.

For an oscillating dipole $\mathbf{p}(t) = \mathbf{p}\mathrm{e}^{-i\omega t}$ placed at $\mathbf{r}_0$, the current density is given by

$$\mathbf{j}(\mathbf{r}, t) = \frac{d\mathbf{p}(t)}{dt}\delta(\mathbf{r} - \mathbf{r}_0) = -i\omega\delta(\mathbf{r} - \mathbf{r}_0)\,\mathbf{p}\,, \tag{B.19}$$

and, by using the continuity equation,

$$i\omega\rho(\mathbf{r}) = \nabla \cdot \mathbf{j}(\mathbf{r})\,, \tag{B.20}$$

the charge density can also be derived. Then, the potentials due to an electric dipole can be obtained by substituting this sources in Eq. (B.16), leading to:

$$\begin{aligned}
\phi(\mathbf{r}) &= \frac{-1}{\epsilon}\int d\mathbf{r}' G_H(|\mathbf{r} - \mathbf{r}'|)\nabla_{\mathbf{r}'}\left[\delta(\mathbf{r}' - \mathbf{r}_0)\mathbf{p}\right] \qquad \text{(B.21)}\\
&= \frac{-1}{\epsilon}\mathbf{p}\cdot\int d\mathbf{r}' G_H(|\mathbf{r} - \mathbf{r}'|)\nabla_{\mathbf{r}'}\delta(\mathbf{r}' - \mathbf{r}_0) = \frac{1}{\epsilon}\mathbf{p}\cdot\int d\mathbf{r}'\delta(\mathbf{r}' - \mathbf{r}_0)\nabla_{\mathbf{r}'}G_H(|\mathbf{r} - \mathbf{r}'|)\\
&= \frac{1}{\epsilon}\mathbf{p}\cdot\nabla_{\mathbf{r}'}G_H(|\mathbf{r} - \mathbf{r}'|)\Big|_{\mathbf{r}'=\mathbf{r}_0} = \frac{-1}{\epsilon}\left(\nabla_{\mathbf{r}}G_H(|\mathbf{r} - \mathbf{r}_0|)\right)\cdot\mathbf{p}\\
\mathbf{A}(\mathbf{r}) &= -\frac{i\omega}{c}G_H(|\mathbf{r} - \mathbf{r}_0|)\mathbf{p}\,.
\end{aligned}$$

By substituting this expressions in Eq. (B.13), the fields can be expressed in the following tensorial form:

$$\mathbf{E}(\mathbf{r}) = \left[\left(\frac{\omega^2}{c^2}\mathbb{1} + \frac{1}{\epsilon}\nabla_{\mathbf{r}}\nabla_{\mathbf{r}}\right)\frac{\mathrm{e}^{ik(|\mathbf{r}-\mathbf{r}_0|)}}{|\mathbf{r} - \mathbf{r}_0|}\right]\cdot\mathbf{p} \equiv \hat{\mathbf{G}}_H(\mathbf{r}, \mathbf{r}_0)\cdot\mathbf{p} \tag{B.22}$$

$$\mathbf{H}(\mathbf{r}) = -\frac{i\omega}{\mu c}\nabla\times\left[G_H(|\mathbf{r} - \mathbf{r}_0|)\mathbf{p}\right] = \left(-\frac{i\omega}{\mu c}\nabla_{\mathbf{r}}G_H(|\mathbf{r} - \mathbf{r}_0|)\right)\times\mathbf{p}\,, \tag{B.23}$$

where $\hat{\mathbf{G}}_H(\mathbf{r}, \mathbf{r}_0)$ stands for Green's tensor which calculates the electric component of the radiation at $\mathbf{r}$ due to a dipole at $\mathbf{r}_0$, in a homogeneous environment. From now on, non-magnetic material will be considered, resulting in wave vectors in each layer given by $k_i = \omega\sqrt{\epsilon_i}/c$. In order to avoid confusions with the wave vector in vacuum, the subscript corresponding to the upper medium will be "up" instead of "0". Thus, the tensor can be rewritten as:

$$\hat{\mathbf{G}}_H(\mathbf{r}, \mathbf{r}_0) = \frac{\omega^2}{c^2}\left(\mathbb{1} + \frac{\nabla_{\mathbf{r}}\nabla_{\mathbf{r}}}{k_{\mathrm{up}}^2}\right)\frac{\mathrm{e}^{ik_{\mathrm{up}}(|\mathbf{r}-\mathbf{r}_0|)}}{|\mathbf{r} - \mathbf{r}_0|}\,. \tag{B.24}$$

Full Green's tensor $\hat{\mathbf{G}}(\mathbf{r}, \mathbf{r}_0)$ will be calculated by including the effect of the substrate. The transfer matrix formalism described in the previous section fulfils the boundary conditions for plane waves with a given $\mathbf{k}$ wave vector. However, the dipole does not emit a plane wave, and it is necessary to decompose the radiation in terms of plane waves. Here, the method used closely follows the one presented in [7], which uses a Fourier transform, converting the spatial dependence of Eq. (B.24) into the (spatial) frequency domain. This is accomplished by using the following substitutions:

$$f(x) = \frac{e^{ik_{up}x}}{x} \Rightarrow \hat{f}(k) = \frac{4\pi}{k^2 - k_{up}^2}$$
$$f(x) = \partial_x g(x) \Rightarrow \hat{f}(k) = ik\hat{f}(k) ,$$

which, together with the linearity of the Fourier transform, lead to:

$$\hat{\mathbf{G}}_H(\mathbf{k}) = \frac{4\pi}{\epsilon_{up}} \left(\frac{k_{up}^2 \mathbb{1} - \mathbf{kk}}{k^2 - k_{up}^2} \right) , \tag{B.25}$$

with $k = |\mathbf{k}|^2$ and $\hat{\mathbf{M}} = \mathbf{kk}$ meaning dyadic product, $(\hat{\mathbf{M}})_{i,j} = k_i k_j$. Therefore, the spatial Green's tensor can be written down using the following integral:

$$\hat{\mathbf{G}}_H(\mathbf{r}, \mathbf{r}_0) = \frac{1}{2\pi^2 \epsilon_{up}} \iiint d\mathbf{k} \left(\frac{k_{up}^2 \mathbb{1} - \mathbf{kk}}{k^2 - k_{up}^2} \right) e^{i\,\mathbf{k}\cdot(\mathbf{r}-\mathbf{r}_0)} . \tag{B.26}$$

Here, the tensor is written in terms of plane waves, but they are not physical, as they admit any wave vector k. This is due to the fact that radiation is an intrinsically two-dimensional phenomenon: the integration along one coordinate will yield the right expression. This integration can be performed using Cauchy's residue theorem. If the integration coordinate chosen is k_z, the integrands have poles $k_z = \pm\sqrt{k_{up}^2 - k_x^2 - k_y^2}$. Outside the real axis, the integral must vanish, and therefore the choice of the upper ($\text{Im}(k_z) > 0$) or lower ($\text{Im}(k_z) < 0$) integration plane depends on the sign of $z - z_0$. Care must be taken in the integration of the $\hat{\mathbf{z}}\hat{\mathbf{z}}$ component, which has a singularity at $\mathbf{r} = \mathbf{r}'$ that must be manually removed. The result of the integration reads:

$$\hat{\mathbf{G}}_H(\mathbf{r}, \mathbf{r}_0) = \frac{i}{2\pi\epsilon_{up}} \iint dk_x dk_y \frac{k_{up}^2 \mathbb{1} - \mathbf{k}_{up}\mathbf{k}_{up}}{k_{up,z}} e^{i\mathbf{k}_{up}\cdot(\mathbf{r}-\mathbf{r}_0)} - \frac{4\pi}{\epsilon_{up}} \hat{\mathbf{z}}\hat{\mathbf{z}}\, \delta(\mathbf{r} - \mathbf{r}_0) , \tag{B.27}$$

with $k_{up,z} = \sqrt{k_{up}^2 - k_x^2 - k_y^2}$ and

$$\mathbf{k}_{up} = k_x \hat{\mathbf{x}} + k_y \hat{\mathbf{y}} + \text{sign}(z_0 - z) k_{up,z} \hat{\mathbf{z}} . \tag{B.28}$$

The plane waves in the integrand of Eq. (B.27) are already bounded to have the wave vector allowed by the medium. In order to replace them with the expressions obtained in Sect. B.1, it is necessary to separate the contributions of the two possible polarizations, s and p. This can be done by introducing a new orthonormal system, given *for each point* (k_x, k_y) by $\hat{\mathbf{k}} \equiv \hat{\mathbf{k}}_{\text{up}}$, $\hat{\mathbf{s}}$ and $\hat{\mathbf{p}}$, as defined in Eq. (B.2). These three vectors form an orthonormal basis, i.e., $\hat{\mathbf{k}}\hat{\mathbf{k}} + \hat{\mathbf{s}}\hat{\mathbf{s}} + \hat{\mathbf{p}}\hat{\mathbf{p}} = \mathbb{1}$. Therefore, Eq. (B.27) can be rewritten as:

$$\hat{\mathbf{G}}_H(\mathbf{r}, \mathbf{r}_0) = \frac{i\omega^2}{2\pi c^2} \iint dk_x dk_y \frac{\hat{\mathbf{s}}\hat{\mathbf{s}} + \hat{\mathbf{p}}\hat{\mathbf{p}}}{k_{\text{up},z}} e^{i\mathbf{k}_{\text{up}}\cdot(\mathbf{r}-\mathbf{r}_0)} - \frac{4\pi}{\epsilon_{\text{up}}} \hat{\mathbf{z}}\hat{\mathbf{z}}\, \delta(\mathbf{r} - \mathbf{r}_0) \,. \tag{B.29}$$

Understanding the meaning of this expression is the key point in the procedure followed. Given a dipole $\mathbf{d}$, the dyadic products $\hat{\mathbf{s}}\hat{\mathbf{s}}$ and $\hat{\mathbf{p}}\hat{\mathbf{p}}$ fulfil $(\hat{\mathbf{s}}\hat{\mathbf{s}})\mathbf{d} = \hat{\mathbf{s}}(\hat{\mathbf{s}} \cdot \mathbf{d})$. Therefore, the amplitude of the plane wave for a particular wave vector is proportional to the projection of the dipole on the polarization directions. The reflection will be obtained by means of the following steps:

1. projecting the dipole in the down-going wave polarizations $\hat{\mathbf{s}}(-k_{\text{up},z})$, $\hat{\mathbf{p}}(-k_{\text{up},z})$
2. adding the dephasing of the wave as it goes from the source to the upper interface
3. multiplying it by the generalized reflection coefficient in Eq. (B.9)
4. adding the dephasing due to position of the measurement point

Thus, the total field at point $\mathbf{r}_0$ will be given by a direct contribution, calculated using Eq. (B.24), and the reflected contribution of the down-going waves, i.e.,

$$\hat{\mathbf{G}}(\mathbf{r}, \mathbf{r}_0) = \hat{\mathbf{G}}_H(\mathbf{r}, \mathbf{r}_0) + \hat{\mathbf{G}}_{\text{refl}}(\mathbf{r}, \mathbf{r}_0) \tag{B.30}$$

with reflected Green's tensor being

$$\begin{aligned}\hat{\mathbf{G}}_{\text{refl}}(\mathbf{r}, \mathbf{r}_0) = \frac{i\omega^2}{2\pi c^2} \iint dk_x dk_y & \frac{\hat{\mathbf{s}}(+k_{\text{up},z})\mathcal{R}^s_{0,N}\hat{\mathbf{s}}(-k_{\text{up},z}) + \hat{\mathbf{p}}(+k_{\text{up},z})\mathcal{R}^p_{0,N}\hat{\mathbf{p}}(-k_{\text{up},z})}{k_{\text{up},z}} \\ & \times \exp\left[i\left(k_x(x - x_0) + k_y(y - y_0) + k_{\text{up},z}(z + z_0)\right)\right] \,.\end{aligned} \tag{B.31}$$

Using polar coordinates, with $k_x = k_\rho \cos k_\phi,\ k_y = k_\rho \sin k_\phi$, the dyadic products are given by:

$$\hat{\mathbf{s}}(+k_{\text{up},z})\hat{\mathbf{s}}(-k_{\text{up},z}) = \begin{pmatrix} \sin^2 k_\phi & -\sin k_\phi \cos k_\phi & 0 \\ -\sin k_\phi \cos k_\phi & \cos^2 k_\phi & 0 \\ 0 & 0 & 0 \end{pmatrix} \tag{B.32}$$

$$\hat{\mathbf{p}}(+k_{\text{up},z})\hat{\mathbf{p}}(-k_{\text{up},z}) = \frac{k_{\text{up},z}}{k^2_{\text{up}}} \begin{pmatrix} -k_{\text{up},z}\cos^2 k_\phi & -k_{\text{up},z}\cos k_\phi \sin k_\phi & -k_\rho \cos k_\phi \\ -k_{\text{up},z}\cos k_\phi \sin k_\phi & -k_{\text{up},z}\sin^2 k_\phi & -k_\rho \sin k_\phi \\ k_\rho \cos k_\phi & k_\rho \sin k_\phi & k^2_\rho / k_{\text{up},z} \end{pmatrix} .$$

As $k_{\text{up},z}$ and the reflection coefficients depend only on k_ρ, the k_ϕ-dependent parts of the integrands are a product of trigonometric functions times an exponential. These integrals define the Bessel functions, and therefore the integration can be performed. For simplicity $x - x_0 \equiv \rho \cos \phi,\ y - y_0 \equiv \rho \sin \phi$, and the result reads

$$\hat{\mathbf{G}}_{\text{refl}}(\mathbf{r},\mathbf{r}_0) = \frac{i\omega^2}{c^2}\int_0^\infty dk_\rho \left[\mathcal{R}^s_{0,N}\hat{\mathbf{f}}^s + \mathcal{R}^p_{0,N}\hat{\mathbf{f}}^p\right] e^{ik_{\text{up},z}(z+z_0)} , \tag{B.33}$$

with the components of the tensors $\hat{\mathbf{f}}^s$, $\hat{\mathbf{f}}^p$ given by

$$\begin{aligned}
f^s_{x,x} &= \frac{1}{k_{\text{up},z}\rho}\left[k_\rho\rho\sin^2(\phi)J_0(k_\rho\rho) + \cos(2\phi)J_1(k_\rho\rho)\right] \\
f^s_{x,y} &= f^s_{y,x} = -\frac{1}{k_{\text{up},z}\rho}\sin(\phi)\cos(\phi)\left[k_\rho\rho J_0(k_\rho\rho) - 2J_1(k_\rho\rho)\right] \\
f^s_{y,y} &= \frac{1}{k_{\text{up},z}\rho}\left[k_\rho\rho\cos^2(\phi)J_0(k_\rho\rho) - \cos(2\phi)J_1(k_\rho\rho)\right] \\
f^s_{x,z} &= f^s_{y,z} = f^s_{z,x} = f^s_{z,y} = f^s_{z,z} = 0
\end{aligned} \tag{B.34}$$

$$\begin{aligned}
f^p_{x,x} &= -\frac{k_{\text{up},z}}{k^2_{\text{up}}\rho}\left(k_\rho\rho\cos^2(\phi)J_0(k_\rho\rho) - \cos(2\phi)J_1(k_\rho\rho)\right) \\
f^p_{x,y} &= f^p_{y,x} = -\frac{k_{\text{up},z}}{k^2_{\text{up}}\rho}\sin(\phi)\cos(\phi)\left[k_\rho\rho J_0(k_\rho\rho) - 2J_1(k_\rho\rho)\right] \\
f^p_{x,z} &= f^p_{z,x} = -\frac{ik^2_\rho}{k^2_{\text{up}}}\cos(\phi)J_1(k_\rho\rho) \\
f^p_{y,y} &= -\frac{k_{\text{up},z}}{k^2_{\text{up}}\rho}\left(k_\rho\rho\sin^2(\phi)J_0(k_\rho\rho) + \cos(2\phi)J_1(k_\rho\rho)\right) \\
f^p_{y,z} &= f^p_{z,y} = -\frac{ik^2_\rho}{k^2_{\text{up}}}\sin(\phi)J_1(k_\rho\rho) \\
f^p_{z,z} &= \frac{k^3_\rho}{k^2_{\text{up}}k_{\text{up},z}}J_0(k_\rho\rho) .
\end{aligned} \tag{B.35}$$

These integrals are the so-called Sommerfeld integrals [5]. By making a spatial rotation in the system, so that $\phi = 0$ and using the symmetries of the system, the number of independent terms can be reduced to 6: $f^s_{x,x}$, $f^s_{y,y}$, $f^p_{x,x}$, $f^p_{x,z}$, $f^p_{y,y}$, $f^p_{z,z}$.

B.3 Numerical Integration of the Sommerfeld Integrals

The numerical evaluation of the integrals defined by Eq. (B.33) is in general complicated, due to a variety of factors:

- Branch cuts: the presence of square roots $\sqrt{k_i^2 - k_\rho^2}$ produces branch cuts in the integrand region, which represent lateral waves ($k_\rho = \pm k_i$). As shown in [4] using symmetry considerations, only the upper and lower media produce branch cuts.

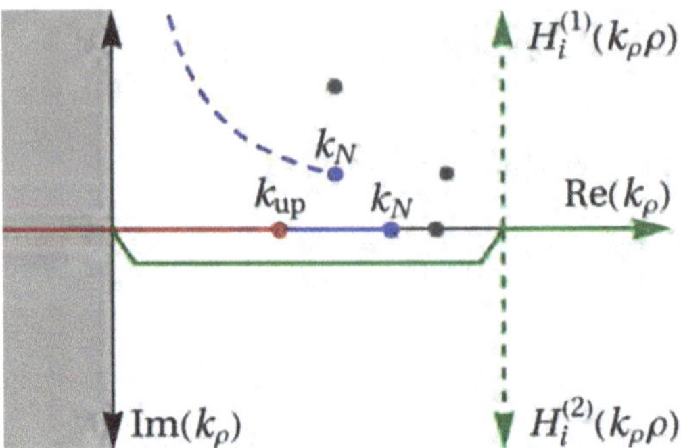

Fig. B.3 Schematics of a possible integration path. The path itself is shown with a *solid green arrow*. A branch cut due to the upper medium (dielectric) is shown in *red*, whereas two different possibilities are shown for the lower one, using *solid blue* (dielectric medium) and *dashed blue* (lossy medium) lines. Some illustrative poles are plotted using *grey disks*. Finally, *dashed green arrows* represent an alternative to the last segment of the integration path, if the Bessel functions are replaced by Hankel functions

As the integration runs over the positive real axis, only two branch cuts will need to be considered, lying at $+k_{\text{up}},\ +k_N$. Physically, they represent lateral waves [8].

- Poles: the generalized reflection coefficient can be expressed as a fraction, whose denominator can vanish for different values of k_ρ. These values represent guided modes supported by the layers of the structure. The best known example is the SPP for TM waves in systems with one interface, for which the generalized reflection coefficient is simply $\mathcal{R}_{0,1}^{p}$ in Eq. (B.1), whose denominator vanishes for

$$\epsilon_{\text{down}} k_{\text{up},z} = -\epsilon_{\text{up}} k_{\text{down},z} \ .$$

- Highly oscillatory integrand: for $k_\rho \rho \gg 1$, the Bessel functions become highly oscillatory.

For passive media ($\mathrm{Im}(k_i) < 0$) and in the region of interest for the integration $\Re(k_\rho) \geq 0$, both the branch points and the poles are in the real axis or in the first quadrant of the complex plane, i.e., they satisfy $\mathrm{Im}(k_\rho) \geq 0$. Therefore, the integration path can be deformed to run in the fourth quadrant and avoid the region of the real axis with branch cuts or poles, and then make it return to the real axis. This is illustrated in Fig. B.3, with the integration path plotted with a solid green arrow.

For measurement points close to the interface, the exponential term $\exp(ik_{\text{up},z}z)$ decays very slowly. If, at the same time, the horizontal distance to the emitter ρ is large, the integrals turn out to be very complicated to evaluate along the real axis. In this situation, it is convenient to rewrite Bessel functions in terms of Hankel functions,

$$H_n^{(1)}(x) = J_n(x) + iY_n(x) \qquad H_n^{(2)}(x) = J_n(x) - iY_n(x) \ , \tag{B.36}$$

with $Y_n(x)$ being Bessel functions of the second kind. The following identity is directly obtained from the definition:

$$J_n(x) = \left(H_n^{(1)}(x) + H_n^{(2)}(x)\right)/2\,. \tag{B.37}$$

The advantage of this substitution is that the asymptotic expansion of the Hankel function is exponential, whereas for Bessel functions it is an inverse square root times a cosine [8]:

$$\begin{aligned}
\lim_{x\to\infty} J_n(x) &\sim \sqrt{\frac{2}{\pi x}}\cos\left(x - \frac{\pi n}{2} - \frac{\pi}{4}\right)\;,\;\mathrm{Im}(x) = 0\\
\lim_{|z|\to\infty} H_n^{(1)}(z) &\sim \sqrt{\frac{2}{\pi z}}\exp\left[i\left(z - \frac{\pi n}{2} - \frac{\pi}{4}\right)\right]\\
\lim_{|z|\to\infty} H_n^{(2)}(z) &\sim \sqrt{\frac{2}{\pi z}}\exp\left[-i\left(z - \frac{\pi n}{2} - \frac{\pi}{4}\right)\right]\,.
\end{aligned} \tag{B.38}$$

Therefore, each integral is split in two parts: the path of the one containing $H_n^{(1)}(k_\rho\rho)$ is deformed to run parallel to the positive imaginary axis, whereas the other, with the terms $H_n^{(1)}(k_\rho\rho)$, runs parallel to the negative imaginary axis (dashed green lines in Fig. B.3).

References

1. A. Sommerfeld, Über die Ausbreitung der Wellen in der drahtlosen Telegraphie. Annalen der Physik **333**, 665–736 (1909)
2. O.J.F. Martin, N.B. Piller, Electromagnetic scattering in polarizable backgrounds. Physical Review E **58**, 3909–3915 (1998)
3. M. Paulus et al., Accurate and eficient computation of the Green's tensor for stratified media. Physical Review E **62**, 5797–5807 (2000)
4. W. Chew Waves and Fields in Inhomogeneous Media. IEEE Press (1999).
5. L. Felsen and N. Marcuvitz Radiation and Scattering of Waves. Wiley (1994).
6. R. P. Álvarez and F. G. Moliner Transfer Matrix, Green Function and Related Techniques: Tools for the Study of Multilayer Heterostructures. Universitat Jaume I. Servei de Comunicació i Publicacions (2004).
7. F.J. García de Abajo, A. Howie, Retarded field calculation of electron energy loss in inhomogeneous dielectrics. Physical Review B **65**, 115418 (2002)
8. M. Abramowitz and I. Stegun Handbook of Mathematical Functions with Formulas, Graphs, and Mathematical Tables. Courier Dover Publications (1964).

Zeitfracht Medien GmbH
Ferdinand-Jühlke-Straße 7
99095 Erfurt, Deutschland
produktsicherheit@kolibri360.de